Stochastics, Algebra and Analysis
in Classical and Quantum Dynamics

Volume 59

Stochastics, Algebra and Analysis in Classical and Quantum Dynamics

Proceedings of the IVth French-German Encounter on Mathematics and Physics, CIRM, Marseille, France, February/March 1988

edited by

S. ALBEVERIO

Fachbereich Mathematik, Ruhr-Universität, Bochum, F.R.G. and BiBoS

PH. BLANCHARD

Fakultät für Physik, Universität Bielefeld, F.R.G. and BiBoS

and

D. TESTARD

Centre de Physique Theorique, CNRS-Luminy, Marseille, and University of Aix-Marseille, France

KLUWER ACADEMIC PUBLISHERS

DORDRECHT / BOSTON / LONDON

Library of Congress Cataloging in Publication Data

French-German Encounter in Mathematics and Physics (4th : 1988 :
Centre national de recontres mathématiques)
Stochastics, algebra, and analysis in classical and quantum
dynamics : proceedings of the IVth French-German Encounter on
Mathematics and Physics, CIRM, Marseille, France, February/March
1988 / edited by S. Albeverio, Ph. Blanchard, and D. Testard.
p. cm. -- (Mathematics and its applications)
ISBN-13: 978-94-011-7978-2 e-ISBN-13: 978-94-011-7976-8
DOI: 10.1007/978-94-011-7976-8
1. Quantum theory--Congresses. 2. Stochastic processes-
-Congresses. 3. Numerical analysis--Congresses. 4. Mathematical
physics--Congresses. I. Albeverio, Sergio. II. Blanchard,
Philippe. III. Testard, D., 1941- . IV. Title. V. Series:
Mathematics and its applications (Kluwer Academic Publishers.
QC173.96.F74 1988
530.1'2--dc20 89-71703

ISBN-13: 978-94-011-7978-2

Published by Kluwer Academic Publishers,
P.O. Box 17, 3300 AA Dordrecht, The Netherlands.

Kluwer Academic Publishers incorporates
the publishing programmes of
D. Reidel, Martinus Nijhoff, Dr W. Junk and MTP Press.

Sold and distributed in the U.S.A. and Canada
by Kluwer Academic Publishers,
101 Philip Drive, Norwell, MA 02061, U.S.A.

In all other countries, sold and distributed
by Kluwer Academic Publishers Group,
P.O. Box 322, 3300 AH Dordrecht, The Netherlands.

Printed on acid-free paper

SERIES EDITOR'S PREFACE

'Et moi, ..., si j'avait su comment en revenir, je n'y serais point allé.'

Jules Verne

The series is divergent; therefore we may be able to do something with it.

O. Heaviside

One service mathematics has rendered the human race. It has put common sense back where it belongs, on the topmost shelf next to the dusty canister labelled 'discarded non-sense'.

Eric T. Bell

Mathematics is a tool for thought. A highly necessary tool in a world where both feedback and non-linearities abound. Similarly, all kinds of parts of mathematics serve as tools for other parts and for other sciences.

Applying a simple rewriting rule to the quote on the right above one finds such statements as: 'One service topology has rendered mathematical physics ...'; 'One service logic has rendered computer science ...'; 'One service category theory has rendered mathematics ...'. All arguably true. And all statements obtainable this way form part of the raison d'être of this series.

This series, *Mathematics and Its Applications*, started in 1977. Now that over one hundred volumes have appeared it seems opportune to reexamine its scope. At the time I wrote

> "Growing specialization and diversification have brought a host of monographs and textbooks on increasingly specialized topics. However, the 'tree' of knowledge of mathematics and related fields does not grow only by putting forth new branches. It also happens, quite often in fact, that branches which were thought to be completely disparate are suddenly seen to be related. Further, the kind and level of sophistication of mathematics applied in various sciences has changed drastically in recent years: measure theory is used (non-trivially) in regional and theoretical economics; algebraic geometry interacts with physics; the Minkowsky lemma, coding theory and the structure of water meet one another in packing and covering theory; quantum fields, crystal defects and mathematical programming profit from homotopy theory; Lie algebras are relevant to filtering; and prediction and electrical engineering can use Stein spaces. And in addition to this there are such new emerging subdisciplines as 'experimental mathematics', 'CFD', 'completely integrable systems', 'chaos, synergetics and large-scale order', which are almost impossible to fit into the existing classification schemes. They draw upon widely different sections of mathematics."

By and large, all this still applies today. It is still true that at first sight mathematics seems rather fragmented and that to find, see, and exploit the deeper underlying interrelations more effort is needed and so are books that can help mathematicians and scientists do so. Accordingly MIA will continue to try to make such books available.

If anything, the description I gave in 1977 is now an understatement. To the examples of interaction areas one should add string theory where Riemann surfaces, algebraic geometry, modular functions, knots, quantum field theory, Kac-Moody algebras, monstrous moonshine (and more) all come together. And to the examples of things which can be usefully applied let me add the topic 'finite geometry'; a combination of words which sounds like it might not even exist, let alone be applicable. And yet it is being applied: to statistics via designs, to radar/sonar detection arrays (via finite projective planes), and to bus connections of VLSI chips (via difference sets). There seems to be no part of (so-called pure) mathematics that is not in immediate danger of being applied. And, accordingly, the applied mathematician needs to be aware of much more. Besides analysis and numerics, the traditional workhorses, he may need all kinds of combinatorics, algebra, probability, and so on.

In addition, the applied scientist needs to cope increasingly with the nonlinear world and the extra mathematical sophistication that this requires. For that is where the rewards are. Linear models are honest and a bit sad and depressing: proportional efforts and results. It is in the non-linear world that infinitesimal inputs may result in macroscopic outputs (or vice versa). To appreciate what I am hinting at: if electronics were linear we would have no fun with transistors and computers; we would have no TV; in fact you would not be reading these lines.

There is also no safety in ignoring such outlandish things as nonstandard analysis, superspace and anticommuting integration, p-adic and ultrametric space. All three have applications in both electrical engineering and physics. Once, complex numbers were equally outlandish, but they frequently proved the shortest path between 'real' results. Similarly, the first two topics named have already provided a number of 'wormhole' paths. There is no telling where all this is leading - fortunately.

Thus the original scope of the series, which for various (sound) reasons now comprises five subseries: white (Japan), yellow (China), red (USSR), blue (Eastern Europe), and green (everything else), still applies. It has been enlarged a bit to include books treating of the tools from one subdiscipline which are used in others. Thus the series still aims at books dealing with:

- a central concept which plays an important role in several different mathematical and/or scientific specialization areas;
- new applications of the results and ideas from one area of scientific endeavour into another;
- influences which the results, problems and concepts of one field of enquiry have, and have had, on the development of another.

There used to be a time that (theoretical) physics and mathematics were practically indistinguishable; then, more recently, there was a period of very little communication between the two subjects, even though, as is now well known, they were on occasion studying the same things (under different names). The example of gauge field theory and the theory of connections and bundles is perhaps the best known example of this.

The interaction between physics and mathematics in this topic continues most fruitfully and dynamically as this volume of 'interaction' papers (in several meanings of the word) on such things as (infinite-dimensional) probability and quantum theory, stochastic dynamics, and cyclic cohomology, amply testifies.

The shortest path between two truths in the real domain passes through the complex domain.

J. Hadamard

La physique ne nous donne pas seulement l'occasion de résoudre des problèmes ... elle nous fait pressentir la solution.

H. Poincaré

Never lend books, for no one ever returns them; the only books I have in my library are books that other folk have lent me.

Anatole France

The function of an expert is not to be more right than other people, but to be wrong for more sophisticated reasons.

David Butler

Bussum, January 1990

Michiel Hazewinkel

CONTENTS

Preface ix

Dedication xv

White Noise and Stochastic Analysis
J. Asch and J. Potthoff 1

Quelques Remarques sur le Cut-Locus sous-Riemannien
G. Ben Arous and R. Léandre 15

Simulation on Random Graphs of the Epidemic Dynamics of Sexually Transmitted Diseases – A New Model for the Epidemiology of AIDS
G. F. Bolz 23

Sur une version à deux dimensions du modele de Ginzburg et Landau pour la supraconductivité
A. Boutet de Monvel 53

Differential Calculus and Integration by Parts on Poisson Space
E. A. Carlen and E. Pardoux 63

Introduction to Entire Cyclic Cohomology (of Z/2-Graded Banach Algebras)
D. Kastler 75

Stochastic Analysis, Mathematical Structures and Distributions Methods
P. Kreé 153

Ward Identities for Conformal Models
S. Lazzarini and R. Stora 169

Dynamical Systems in Plasma Theory
E. K. Maschke 173

Bosonic Strings and Measures on Infinite Dimensional Manifolds
S. Paycha 189

A Reaction-Diffusion Model for Moderately Interacting Particles
H. Rost 205

Nondegeneracy in the Perturbation Theory of Integrable Dynamical Systems
H. Rüssmann 211

Energy Forms in Terms of White Noise
L. Streit 225

Linear Stability and the Parameter Modulation Approach for Solitary Waves in Hamiltonian Systems with Symmetries
J. Stubbe 235

Index 247

PREFACE

The present volume collects the Proceedings of the Ivth French-German-Encounters in Mathematics and Physics that took place at the Centre International de Recherches Mathématiques (CIRM) of the Société Mathématique de France in Luminy, February 28 to March 4, 1988.

This Meeting was organized in the same spirit of the other Encounters in 1981, 1983,. 1985 and as well in the tradition of a very close French-German collaboration under the auspices of a treaty between the University of Marseille-Luminy and the Physics Department of Bielefeld University. This collaboration began in the academic year 1975-1976 during the research year "Mathematical Problems of Quantum Dynamics" at the Centre for Interdisciplinary Research (ZiF) of Bielefeld University. French and German mathematicians and physicists suggested that CIRM and ZiF might serve as a "pair of houses" for French-German scientific exchange on the interdisciplinary interface between Mathematics and Physics.

Theoretical Physics is indeed the best example of the way Mathematics and Physics are complementary. New areas of Mathematics have been created and developed from attempts to understand the laws of Physics. Conversely Mathematics has greatly helped new developments in Physics. Let us mention for example the impact of Group Theory in connection with the understanding of the symmetries of nature, the interactions between Differential Geometry and Physics (General Theory of Relativity, Non-Abelian Gauge Theory ....), and the role of Probability Theory in constructing quantum fields and conversely the stimulation of new developments in Stochastic Analysis through the study of Quantum Theory.

From the point of view of Functional Integration, Quantum Field Theory can be viewed as the study of probability measures associated with classical mechanics, which implies that phenomena occurring in statistical physics-like phase transitions and symmetry breaking - are also present in Quantum Field Theory and conversely. During the past decades the boundaries between many domains of Mathematics and Theoretical Physics have practically disapeared.

A look at the table of contents of this volume is a good illustration of this very fruitful interaction and reveals that the papers appearing here cover a number of different aspects of Probability Theory, Quantum Field Theory, the theory of Dynamical Systems, Statistical Physics and Cyclic Cohomology.

Also, the lectures point out a strong interaction between different subjects: Let us mention the presence and fertility of relations between Probability Theory and Quantum Field Theory and the possible intriguing connections between Cyclic Cohomology and the study of relativistic quantum fields. Let us examine shortly the various contributions to this volume.

H. Rost considers a class of models describing "moderate interaction" of n independently moving particles in $\mathbb{R}^d$ and shows that in a suitable limit as $n \to +\infty$ a certain non-linear reaction diffusion equation is obtained, which is independent of the scaling sequence.

The contributions of J. Asch, J. Potthoff and L. Streit are connected to what is called White Noise Analysis. This infinite dimensional analysis was proposed by T. Hida to analyse functionals of Brownian Motion by using White Noise. The aim of the paper by J. Asch and J. Potthoff is to describe the fundamental notions of White Noise Analysis, to develop its stochastic calculus and a recent extension of it to non-anticipating functionals.

The contribution of L. Streit describes the construction of a special class of Dirichlet forms, the so called energy forms, in infinitely many dimensions in the framework of White Noise Analysis. For this a detailed account of spaces of test functions and generalized functionals of White Noise is given. The problem of the closability of the form is also discussed.

The contribution of Ben Arous and Léandre is concerned with the problem of computing an asymptotic expansion, valid for small time, for the heat kernel associated to a class of hypoelliptic operators. The authors prove a conjecture by Bismut which states that, outside a well characterized set of pairs of points (x,y) (the under-riemannian cut-locus), the heat kernel $p_t(x,y)$ has the same form as in the elliptic case, for small time. Counterexamples are described, which exhibit different types of estimations in the cut locus and which suggest the necessity of a different approach for pair of points in this set.

The main goal of the contribution by E. Carlen and E. Pardoux is to develop a calculus playing on standard Poisson space the same role as the differential calculus on Wiener space. Directional derivatives, the gradient and divergence operators and a Malliavin calculus are introduced and discussed on Poisson space.

It is always difficult and open to controversy to try to assess the development of recent domains of research. Stochastic Analysis being by its nature an interdisciplinary undertaking, wide open to applications and having had an enormous development in recent decades, allows especially for very different points of view concerning its "very nature". P. Krée in his contribution takes the opportunity to present a personal historical look at the development of Stochastic Analysis in relation to physical applications, stressing the importance of distributional methods in Probability Theory. Obviously very different points of view can be hold (and are indeed associated with other specialists in the field).

The aim of the contribution of G.F. Bolz is to explain some results about a stochastic model describing the dynamics of AIDS-epidemics and more generally of sexually transmissible diseases as a discrete stochastic Markov process on random graphs. This approach to the Mathematical Epidemiology of sexually transmitted diseases is new in linking the structure of the random graph and the dynamics of the epidemics to social and medical data as well as in exploiting the richness of the structure of stochastic processes over random graphs to model a very complex epidemiological situation. G.F. Bolz shows how the standard modelling of epidemics using systems of ordinary differential equation appears as a special limiting case of the stochastic model on random graphs. Moreover he describes computational aspects of this framework and presents first numerical results.

A. Boutet de Monvel applies to a superconductor of cylindrical shape a model proposed in 1950 by Ginsburg and Landau. Due to the symmetry the related boundary value problem is two-dimensional. The author shows that the configuration space is not connected. Moreover she proves that the extremal solutions are regular and in a special case she is able to construct explicit solutions which minimize in each connected component the corresponding energy functional.

The contribution of Lazzarini and Stora concerns the research of Ward identities in the context of conformally invariant models. Different proposals concerning the case of free bosonic strings are studied with a particular interest with respect to their general validity. Geometric arguments are presented which are in favour of Becchi's version of Ward identities.

The contribution of S. Paycha is devoted to a new approach to the quantization problem of Bosonic strings. It is shown that the by now classical Polyakov measure can be partially reduced using a convenient rigourous treatment, to the measure which appears in two dimensional models with an exponential interaction (the Liouville model measure) whose coupling constant depends on the dimension of the space of evolution of strings. The main point in this approach is that it allows to treat the case of low dimensions, less or equal to 13, and different from the critical $d = 26$ obtained in the conventional approach. The price to be paid is the fact that an integration with respect to the Liouville measure is required. It turns out that this is still tractable and, if one is able to treat the Teichmüller parameters dependence of the Liouville measure, one can expect to get information and estimates for the physical quantities.

The important contribution of D. Kastler to this volume is devoted to the extension to the case of $\mathbb{Z}/2$-graded Banach algebras of Conne's Entire Cyclic Cohomology Theory. As one knows, this approach, motivated at the beginning by the Atiyah-Singer's notion of a global elliptic operator, has already proved its efficiency and revealed itself as a fundamental tool: among results coming from the techniques involved let us mention the recent proof of the Novikov's conjecture for hyperbolic groups. The contribution in this volume is an extensive self-contained account of results with very detailed proofs. As mentioned before and motivated by supersymmetric Quantum Field Theory, the $\mathbb{Z}/2$-graded version of the theory is presented. It is shown, for instance, that cyclic cohomology classes of arbitrary order appear naturally in a expansion related to a supersymmetric K.M.S. condition, a result which seems to be important for the physical status of the theory.

In the recent years, in connection with quantization of non-integrable classical systems, the study of quasi periodic or stochastic Hamiltonians in Quantum Physics or the investigation concerning the Quantum Chaos problematics, techniques of the theory of Dynamical System have played an increasing role in Mathematical Physics. This volume gathers three contributions to this subject.

The first contribution by H. Rüssmann is concerned with a purely classical problem, precisely the problem of stability in Celestial Mechanics. Let us recall that the fundamental Kolmogorov, Arnold, Moser theorem (KAM) has given about thirty years ago the proof of existence of invariant tori for systems which are obtained by a small smooth perturbation of an integrable Hamiltonian. Among the hypothesis of the KAM theorem, there is the condition that the unperturbed Hamiltonian has to fulfill a strong non-degeneracy condition, which unfortunately, does not allow to apply the theorem to the area which gave the initial motivation for the study, namely to handle the problem of stability of planetary orbits. Rüssmann formulates a new condition of non-degeneracy which is optimal from a mathematical point of view, allows to treat the orbit problem and requires, in turn, to develop a new method of proof for KAM-like theorems.

In the part of this volume devoted to Dynamical Systems, another contribution by J. Stubbe concerns the stability of soliton-like solutions for a specific class of Hamiltonian systems. More precisely, he is interested in the linear stability of soliton solutions of non linear wave equations. As a main result, J. Stubbe, proves that the soliton character of a solution is preserved under a small perturbation whose only effect is the modulation of the solitary wave parameters.

E. Maschke gives an introduction to situations encountered in experimental plasma physics which can be described by Dynamical Systems. Namely he studies the plasma confinement by a magnetic field in a Tokamak, as well as magnetohydrodynamical instabilities, in connection with nuclear fusion. An important part of the paper is devoted to a relatively simple modelization giving equations identical with those associated with a two-dimensional Bénard convection and are convenient to get a good account of experimental results. Such typical properties of Dynamical Systems as appearence of strange attractors, intermittency, bifurcation cascades, are numerically investigated and compared with experiments.

It was of great help to have support of the authorities of the "Conseil Municipal" of Marseilles and particularly of the "Comité du Jumelage Marseille-Hamburg". Moreover we gratefully acknowledge the support of the "Centre National pour la Recherche Scientifique", the Universities of Aix-Marseille I and Aix-Marseille II, the "Conseil Regional de la Région Provence-Alpes-Côte d'Azur", the latter for helping us to publish this book.

We are also grateful to the staff of CIRM and CPT, in particular to A. Zeller-Meier, M. Cohen-Solal and N. Lambert, who expertly handled the organization of the meeting.

Last but not least we would like to thank Michiel Hazewinkel for helping us with the publication of the book.

S. Albeverio   Ph. Blanchard   D. Testard

September 1989

RAPHAEL HOEGH-KROHN

MICHEL SIRUGUE

DEDICATION

This book is dedicated to the very dear memory of two friends, Raphael Hoegh-Krohn and Michel Sirugue. Fate has decreed that we should lose them in the space of just over two months. 11th November 1987, Michel succumbed to an implacable illness against which he had struggled until the very last with courage and lucidity. 24th January 1988, the news of the death of Raphael reached us from Norway, like a thunderbolt.

It is with some trepidation that we write a few lines about them in the proceedings of a Conference in the preparations for which Michel took a very active part. For in such circumstances, there is always the risk of saying too much and not enough at the same time.

While they were in many ways very different, they both had in common what is essential, namely intelligence and warmth of heart. These qualities of mind and feeling showed themselves in their generosity, their accessibility, and their concern for others - particularly those who most needed it, like young people and beginners. Rather than occupying a position, they had an audience. For both of them it was not enough merely to give what they had, they gave what they were: and this will never be forgotten by their friends. Their common aim was to think clearly, to speak honestly and to act fairly. Both devoted themselves to what was important without ever yielding to self-importance. They considered humour and cheerfulness as necessary conditions for open-mindedness, and, in order to be capable of doing great things, they were able to resist the temptation of devoting themselves too much to small things. To continue in the path they have shown, try to work in their open spirit, is the sole consolation we have. As was strongly felt during the conference and is also apparent from the proceedings, the scientific and human trace of these great men and scientists will accompany us steadily.

WHITE NOISE AND STOCHASTIC ANALYSIS

J. Asch and J. Potthoff
Fachbereich Mathematik
Technische Universität Berlin
D-1000 Berlin 12

ABSTRACT. In this article we describe the fundamental notions of T. Hida's white noise calculus and show how they can be used for an approach of stochastic analysis with a more functional-analytic flavour. In particular, we shall review the construction of stochastic integrals (more general than in Itô's calculus) of Kuo and Russek [KR 88] and show that Itô's lemma holds (in its usual form) for these integrals.

1. INTRODUCTION

In 1975 T. Hida proposed to analyze functionals of Brownian motion with the help of white noise [Hi 75]. The basis of their relation is simply that informally speaking white noise $x(t)$, $t \in \mathbb{R}$, is the time derivative $\dot{B}(t)$ of Brownian motion B at time t. More formally, we can express (a version of) Brownian motion as the process $(B(t), t \in \mathbb{R}_+)$ given by the random variables $\langle x, 1_{[0,t)} \rangle$ on the white noise space $(S^*(\mathbb{R}), \mathcal{B}, d\mu)$, $d\mu$ being the white noise measure (cf. section 2).
The advantage of this point of view, i.e. taking white noise as the basic object, is that it is simpler than Brownian motion. For example, it is "independent at different moments of time":

$$\mathbb{E}\, x(t)x(s) = \delta(t-s) \tag{1.1}$$

However, x is no longer an ordinary stochastic process, but a generalized one.
The differential calculus based on white noise proposed in [Hi 75] and the idea [Hi 75,80] to define suitable spaces of test and generalized functionals of white noise has since been taken up and developped by several authors, cf. e.g. [KT 80, Kuo 83] and literature quoted there. Among other domains the white noise calculus found applications in quantum dynamics [HS 77], in the theory of Dirichlet forms in infinite dimensions [HPS 88a, HPS 88b, Str 88] in turbulence theory [Ch 86] and in stochastic analysis [Kub 83, KR 88, KS 87, Po 87 , Tak 83].
In this article we review the papers [KR 88, AP 88] which treat

S. Albeverio et al. (eds.), Stochastics, Algebra and Analysis in Classical and Quantum Dynamics, 1–14.

stochastic integrals and their calculus in the white noise set-up. We shall see that Hida's idea to use "sharp time" functional derivatives with respect to white noise is quite essential for a neat theory. The stochastic integral we define does not make use of non-anticipatory conditions and is an extension of Ito's stochastic integral for non-anticipating inegrands. However, it is a different extension than the Skorohod integral (cf. e.g. [NP 88, PP 87]). The plan of this article is as follows. In section 2 we review the basic ingredients of the white noise calculus. Stochastic integrals and Itô's lemma will be discussed in section 3.

2. WHITE NOISE CALCULUS

Consider the Schwartz space $\mathcal{S}^*(\mathbb{R})$ of tempered distributions and let $\mathcal{B}$ denote the σ-algebra over $\mathcal{S}^*(\mathbb{R})$ generated by its cylinder sets (weak-∗ open sets). By Minlos' theorem (see e.g. [Hi 70]) we know that there exists a probability measure $d\mu$ on $\mathcal{B}$, so that for all $\xi \in \mathcal{S}(\mathbb{R})$

$$\int_{\mathcal{S}^*(\mathbb{R})} e^{i\langle x,\xi\rangle} d\mu(x) = e^{-1/2\|\xi\|_2^2} \tag{2.1}$$

where $\|\cdot\|_2$ is the norm of $L^2(\mathbb{R})$ and $\langle\cdot,\cdot\rangle$ denotes dual pairing between $\mathcal{S}^*(\mathbb{R})$ and $\mathcal{S}(\mathbb{R})$. The associated coordinate process $x: \xi \to \langle x,\xi\rangle$, $\xi \in \mathcal{S}(\mathbb{R})$, is a generalized stochastic process over $\mathcal{S}(\mathbb{R})$ with mean zero and covariance $\text{cov}(\xi,\eta) = (\xi,\eta)$, where $(\cdot,\cdot)$ is the scalar product of $L^2(\mathbb{R})$. This process is called white noise and we shall write informally $x(t)$, $t \in \mathbb{R}$, etc.

Let us denote $L^2(\mathcal{S}^*,\mathcal{B},d\mu)$ by (L^2). Due to Wiener, Itô, Segal, ... it is well-known (cf. e.g. [Hi 80]) that (L^2) has a decomposition:

$$(L^2) = \bigoplus_{n=0}^{\infty} \mathcal{H}^{(n)} \tag{2.2a}$$

$$\mathcal{H}^{(n)} \simeq L^2(\hat{\mathbb{R}^n}, n!\,dt) \tag{2.2b}$$

The right hand side of (2.2b) is the subspace of (a.e.) symmetric functions of $L^2(\mathbb{R}^n)$ and the weight $n!$ is chosen such that (2.2) is an isometric isomorphism of (L^2) and the symmetric Fock space

$$(L^2) \simeq \hat{\mathcal{F}} := \bigoplus_{n=0}^{\infty} L^2(\hat{\mathbb{R}^n}, n!\,dt) \tag{2.3}$$

over $L^2(\mathbb{R})$ (cf. also e.g. [Ne 73, Si 74]). $\mathcal{H}^{(n)}$ is often called "homogeneous chaos-" or "multiple Wiener integral of degree n". The following is a very useful representation of $H^{(n)}$. Each $\varphi^{(n)} \in H^{(n)}$, $n \in \mathbb{N}$, can be written as $(x \in S^*(\mathbb{R}))$

$$\varphi^{(n)}(x) = \int_{\mathbb{R}^n} f^{(n)}(u) :x^{\otimes n}:(u)\,du \tag{2.4}$$

with $f^{(n)} \in L^2(\hat{\mathbb{R}}^n)$ and $:x^{\otimes n}:$ is the "Wick-ordering" of $x^{\otimes n}$. It can be defined recursively as follows. For $u \in \mathbb{R}^n$ set

$$:x^{\otimes n}:(u) = :x(u_1)\ldots x(u_n): \tag{2.5a}$$

$$:1: = 1 \tag{2.5b}$$

$$:x(u): = x(u) \tag{2.5c}$$

$$:x(u_1)\ldots x(u_n): = :x(u_1)\ldots x(u_{n-1}):x(u_n) - \sum_{k=1}^{n-1} \delta(u_n-u_k):x(u_1)\ldots \not{x}(u_k)\ldots x(u_{n-1}): \tag{2.5d}$$

where $\not{x}(u_k)$ means that this factor is omitted.

Example:

$$:x^{\otimes 2}:(u_1,u_2) = x(u_1)x(u_2) - \delta(u_1-u_2) \tag{2.6}$$

It is not hard now to give (2.4) a formal meaning by interpreting the "integral" as dual pairing for $f^{(n)} \in S(\mathbb{R}^n)$ and in general as an (L^2)-limit of such. Note that

$$\|\varphi^{(n)}\|_2^2 = n!\,\|f^{(n)}\|^2_{L^2(\mathbb{R}^n)} \tag{2.7}$$

if $\varphi^{(n)}$ is represented as in (2.4).
By an induction on (2.5) one proves the following result (the "reodering formula") for the (pointwise) product of $\varphi^{(n)}$ given as in (2.4) and $\psi^{(m)}$ with

$$\psi^{(m)}(x) = \int_{\mathbb{R}^m} g^{(n)}(u) :x^{\otimes m}:(u)\,du \tag{2.8}$$

$$(\varphi^{(n)}\psi^{(m)})(x) = \sum_{k=0}^{n\wedge m} k!\binom{n}{k}\binom{m}{k}\int_{\mathbb{R}^{n-k}}\int_{\mathbb{R}^{m-k}}\left(\int_{\mathbb{R}^k} f(s;u)g(s;v)ds\right)\cdot \tag{2.9}$$
$$\cdot :x(u_1)\ldots x(u_{n-k})x(v_1)\ldots x(v_{m-k}):dvdu$$

Now we introduce the S-transformation on (L^2) (cf. [KT 80]). Let $\varphi \in (L^2)$ and $\xi \in S(\mathbb{R})$. Set

$$(S\varphi)(\xi) := \int \varphi(x+\xi)d\mu(x) \tag{2.10}$$

and note the formula ([KT 80])

$$(S\varphi)(\xi) = e^{-1/2\|\xi\|_2^2}\int e^{<x,\xi>}\varphi(x)d\mu(x) \tag{2.11}$$

One easily shows that, if $\varphi^{(n)}$ is given by (2.4) then

$$(S\varphi^{(n)}(\xi) = \int_{\mathbb{R}^n} f^{(n)}(u)\xi^{\otimes n}(u)du \tag{2.12}$$

Hence, if we identify $f^{(n)} \in L^2(\mathbb{R}^n)^{\wedge}$ with its isomorphic image in $(L^2(\mathbb{R}))^{\hat{\otimes} n}$ (n-fold symmetric tensor product), the right hand side of (2.12) is the evaluation of the tensor $f^{(n)}$ on $(\xi,\ldots,\xi) \in L^2(\mathbb{R})^{\times n}$, i.e. S implements (2.3).

Consider again the representation (2.4)

$$\varphi(x) = \sum_{n=0}^{\infty}\int_{\mathbb{R}^n} f^{(n)}(u):x^{\otimes n}:(u)du \tag{2.13}$$

of $\varphi \in (L^2)$ $(f^{(0)} \in \mathbb{C})$. For many purposes it is convenient to introduce subspaces of (L^2) which are such that the sequence $(f^{(n)}; n \in \mathbb{N}_0)$ in (2.13) has "nice" properties (cf. e.g. [AP 88, Hi 80, Hi 85, HPS 88, KT 80, Ku 83, Po 87b, Yo 87]). The basic possibilities of defining such subspaces are:

(a) to require the functions $f^{(n)}$ to belong to a class of "nice" functions, such as Sobolev spaces, weighted Sobolev spaces, test function spaces etc.

(b) to require a certain decrease of the sequence $(|||f^{(n)}|||;$

$n \in \mathbb{N}_0$) where $|||\cdot|||$ is some norm (or a family of (semi-) norms, typically chosen in (a))

Such constructions can be done neatly with the help of so-called second quantized opertors and we refer the reader to the above quoted papers for more details. There is one space $(S) \subset (L^2)$ (whose construction is for example given in [HPS 88, Str 88]) which has the following properties:

- (S) is a nuclear Fréchet space
- (S) is an algebra
- (S) is stable under $\exp(\tau N)$, $\tau \in \mathbb{R}$, N being the number operator: $N\varphi^{(n)} = n\varphi^{(n)}$
- if $\varphi \in (S)$ is represented in (2.13), then $f^{(n)} \in S(\hat{\mathbb{R}}^n)$

So the Gel'fand triple

$$(S)^* \supset (L^2) \supset (S) \tag{2.14}$$

has much in common with the triple

$$S^*(\mathbb{R}) \supset L^2(\mathbb{R}) \supset S(\mathbb{R}) \tag{2.15}$$

and we call elements in (S) white noise test functionals, while those in $(S)^*$ generalized white noise functionals.

On (S) we can conveniently introduce the Hida derivatives ∂_t, $t \in \mathbb{R}$. For $\varphi \in (S)$ and $t \in \mathbb{R}$ set

$$\partial_t\varphi(x) := S^{-1} \frac{\delta}{\delta\xi(t)} S\varphi(x) \tag{2.16}$$

where $\frac{\delta}{\delta\xi(t)}$ is the usual Fréchet functional derivative ((2.16) is well-defined for $\varphi \in (S)$). It is easy to see that $\partial_t : (S) \to (S)$. Let ∂_t^* denote the dual of ∂_t. Then you find the following table of rules (cf. also [Hi 85, KT 80, Po 87a]):

- ∂_t is a derivation on (S)
- ∂_t admits a chain rule
- $\partial_t :x(u_1)\ldots x(u_n): = \sum_{k=1}^{n} \delta(t-u_k) :x(u_1)\ldots \not{x}(u_k)\ldots x(u_n):$
- $\partial_t^* :x(u_1)\ldots x(u_n): = :x(t)x(u_1)\ldots x(u_n):$
- $[\partial_t, \partial_s^*] = \delta(t-s)$, $[\partial_t, \partial_s] = [\partial_t^*, \partial_s^*] = 0$

- $N = \int dt\, \partial_t^* \partial_t$
- $x(t) \cdot = (\partial_t^* + \partial_t)$

3. STOCHASTIC INTEGRAL IN WHITE NOISE CALCULUS

This section is essentially a report on the articles [AP 88, KR 88], where stochastic integrals and their calculus in the framework presented in section 2 is investigated.

Let φ be a process, say over $[0,1]$, in (L^2) and assume that φ is non-anticipatory with respect to Brownian motion $B(t;x) = \langle x, 1_{[0,t)}\rangle$, $t \in \mathbb{R}_+$. Consider Itô's stochastic integral with respect to Brownian motion and the following informal computation:

$$\int_0^1 \varphi(s)dB(s) = \int_0^1 \varphi(s)\dot{B}(s)ds = \int_0^1 x(s)\varphi(s)ds = \int_0^1 (\partial_s^* + \partial_s)\varphi(s)ds \tag{3.1}$$

Here we used that the time derivative $\dot{B}(s)$ of Brownian motion at time s is equal to $x(s)$ and that — according to our table of rules at the end of section 2 — multiplication by $x(s)$ is given by $(\partial_s^* + \partial_s)$. If we want to make (3.1) as the <u>definition</u> of a stochastic integral, we have to consider functionals $\varphi(s)$ of white noise belonging to larger spaces than (S): for example $B(s)$ does not belong to (S). But then one has to face the difficulty that multiplication by $x(s)$, as described in (3.1), is no longer well-defined. To see this, let $\varphi(s;x) = B(s;x) = \langle x, 1_{[0,s)}\rangle$ and $\varphi_1(s,x) = \langle x, 1_{[0,s]}\rangle$. Note that φ_1 too is a Brownian motion. It is plain to compute

$$\partial_s \varphi(s) = 0, \quad \partial_s \varphi_1(s) = 1 \tag{3.2}$$

$$\int_0^1 \partial_s^* \varphi_2(s)ds = \frac{1}{2}(\varphi_2(1)^2 - 1) \tag{3.3}$$

with $\varphi_2 = \varphi_1$ or φ. Therefore two different versions of Brownian motion give different stochastic integrals (3.1). Comparison with Itô's integral suggests that we define [KR 88]

$$\int_0^1 (\partial_s^* + \partial_{s+})\varphi(s)\,ds \tag{3.4}$$

$$\partial_{s+}\varphi(s) := \lim_{\varepsilon \downarrow 0} \partial_{s+\varepsilon}\varphi(s) \tag{3.5}$$

as the white noise version of stochastic integrals. In fact in [KR 88] it is proved that for non-anticipatory φ, $\partial_{s+}\varphi(s) = 0$ for all s and that (3.4) coincides with Itô's integral of φ. However, note that (3.4), (3.5) do not make use of any adaptness condition of φ and therefore present an extension of Itô's stochastic integral to integrands which are not necessarily non-anticipatory. It is worthwhile to point out, that (3.4) is a different extension of Itô's integral than the Skorohod integral, cf. e.g. [NP 88, PP 87]. In fact the latter corresponds to the first term of (3.4).

Let us sketch, how Kuo and Russek treat (3.5) [KR 88] (cf. also [AP 88]). Consider the decomposition (2.13) of $\varphi(s)$, $s \in [0,1]$. Since

$$\partial_t\varphi(s;x) = \sum_{n=1}^{\infty} n \int_{\mathbb{R}^{n-1}} f^{(n)}(s;t,u) : x^{\otimes(n-1)} : (u)\,du \tag{3.6}$$

our problem is essentially to make sense out of

$$\lim_{\varepsilon \downarrow 0} f^{(n)}(s;s+\varepsilon,\cdot) \tag{3.7}$$

in L^2-sense in s and the remaining variables. The existence of the limit (3.7) (in $L^2([0,1]\times\mathbb{R}^{n-1})$) follows then by an application of the trace theorem (cf. e.g. [Ad 75]), if $f^{(n)}$ belongs to the subspace $H^{(n)}_{\alpha,\delta}([0,1]\times\mathbb{R}^n)$ of $L^2([0,1];L^{2\wedge}(\mathbb{R}^n))$ consisting of those elements which are finite in the norm whose square is equal to

$$\| f^{(n)} \|^2_{L^2([0,1];L^{2\wedge}(\mathbb{R}^n))} + \| 1_\delta f^{(n)} \|^2_{H_\alpha(\Omega_\delta^{(n)})} \tag{3.8}$$

for $\alpha > 1/2$. Here

$$\Omega_\delta^{(n)} := \{(s;u) \in (0,1)\times\mathbb{R}^n;\ s < u_1 < s+\delta\} \tag{3.9}$$

and $H_\alpha(\Omega_\delta^{(n)})$ is the standard Sobolev space of order α over the region $\Omega_\delta^{(n)}$. Moreover 1_δ denotes the indicator of $\Omega_\delta^{(n)}$. From the space $H_{\alpha,\delta}^{(n)}$ one now forms a space of Fock type by taking their direct sum (weighted by n!) over $n \in \mathbb{N}_0$ and by (2.3) one defines this way a subsapce $\mathcal{H}_{\alpha,\delta}$ of (L^2). If then $N^{1/2}\varphi \in \mathcal{H}_{\alpha,\delta}$ with $\alpha > 1/2$, $\delta > 0$, it is shown in [KR 88] that the limit (3.5) belongs to $L^2([0,1]; (L^2))$ and that the stochastic integral (3.4) of φ exists in (L^2).

It should be noted, that one can also define in a similar way spaces of functionals φ so that $\partial_{s-}\varphi(s) = \lim_{\varepsilon\uparrow 0} \partial_{s+\varepsilon}\varphi(s)$ exists in the above sense and one can define the corresponding backward integral

$$\int_0^1 (\partial_s^* + \partial_{s-})\varphi(s)ds \tag{3.10}$$

In particular

$$\int_0^1 (\partial_s^* + 1/2(\partial_{s+} + \partial_{s-}))\varphi(s)ds \tag{3.11}$$

is an extension of the Stratonovich integral.

For simplicity let us suppose from now on that for some $\tau > 0$ $\exp(\tau N)\varphi \in \mathcal{H}_{\alpha,\delta}$, $\alpha > 1/2$, $\delta > 1$, and that $\exp(\tau N)\varphi: [0,1] \to (L^2)$, $s \mapsto \exp(\tau N)\varphi(s)$ is continuous. Let us denote

$$X(t) := \int_0^t (\partial_s^* + \partial_{s+})\varphi(s)ds, \quad t \in [0,1] \tag{3.12}$$

We shall be interested to integrate another process ψ with respect to X and define

$$\int_0^t \psi(s)dX(s+) := \int_0^t (\partial_s^* + \partial_{s+})\psi(s)\varphi(s)ds \tag{3.13}$$

whenever the right hand side makes sense. In [AP 88] it is proved, that this is the case, if ψ satisfies the same conditions as φ above with suitably large τ. Moreover if $\alpha > 1$, then the stochastic integral (3.13) is approximated in (L^2) by Riemannian sums:

$$\sum_k \psi(s_k)(X(s_{k+1})-X(s_k)) \xrightarrow{(L^2)} \int_0^t \psi(s)dX(s+) \tag{3.14}$$

where the (s_k) form a partition of $[0,t]$ whose mesh tends to zero.

It is now natural to ask whether there is a lemma of Itô type for our extended stochastic integral (3.4) (and therefore also for (3.13)). The answer is given in [AP 88] and is very simple: Itô's lemma holds in its usual form for (3.4), i.e. if $\varphi_1,\ldots,\varphi_n$ is a collection of processes as above (with $\alpha > 1$, $\delta > 1$) and if F is a C^3-function on $\mathbb{R}^n$, polynomially bounded together with its first three derivatives (and if τ is large enough, depending on the degree of the majorizing polynomials) then (in (L^2)-sense)

$$F(X(t)) - F(X(s)) = \int_s^t \sum_{j=1}^n D_jF(X(u))dX_j(u+) + \frac{1}{2}\int_s^t \sum_{j,k=1}^n D_jD_kF(X(u))\varphi_j(u)\varphi_k(u)du \tag{3.15}$$

where $X = (X_1,\ldots,X_n)$ and X_j, $1 \leq j \leq n$, is the stochastic integral (3.12) of φ_j.

In the remainder of this paper we shall discuss the proof of (3.15). For simplicity we assume that $n = 1$, $F \in C_b^3(\mathbb{R})$ and that φ has only one homogeneous component, namely in the first chaos

$$\varphi(s;x) = \int f(s;u)x(u)du \tag{3.16}$$

The idea of the proof of (3.15) is quite simple: We write

$$F(X(t)) - F(X(s)) = \sum_{k=0}^{N-1} (F(X(s_{k+1})) - F(X(s_k))) \tag{3.17}$$

with $s = s_0 < s_1 <\cdots< s_N = t$. Now we expand the differences on the right hand side

$$F(X(s_{k+1})) - F(X(s_k)) = F'(X(s_k))\Delta_k X + \frac{1}{2}F''(X(s_k))(\Delta_k X)^2 + O((\Delta_k X)^3) \tag{3.18}$$

with

$$\Delta_k X = X(s_{k+1}) - X(s_k) \tag{3.19}$$

Note that by (3.14) we know that the first order term inserted into (3.17) converges in (L^2) to

$$\int_s^t F'(X(u))dX(u+)$$

Consider now the second order term. First we shall show that ($\Delta := [t,t+\varepsilon)$)

$$\varepsilon^{-1}(\Delta X)^2 \rightarrow \varphi(t)^2 \tag{3.20}$$

weakly in (L^2) as $\varepsilon \to 0$, uniformly in $t \in [0,1)$. (From the arguments leading to (3.20) it will be clear that the third order term will not contribute in the limit and henceforth be ignored.)
With the rules given in section 2 one finds immediately

$$X(t+\varepsilon;x) - X(t;x) = g_\Delta^{(o)} + \int_{\mathbb{R}^2} g_\Delta^{(2)}(u):x^{\otimes 2}:(u)du \tag{3.21}$$

with

$$g_\Delta^{(o)} = \int_\Delta f(s;s+)ds \tag{3.22}$$

$$g_\Delta^{(2)}(u_1,u_2) = 1_\Delta(u_1)f(u_1;u_2) \tag{3.23}$$

(For convenience we did not symmetrize $g_\Delta^{(2)}$). Using the relations (2.5) we find

$$\begin{aligned}(\Delta X)^2(x) = \int_\Delta \varphi(u;x)^2 du + h_\Delta^{(o)} + \int_{\mathbb{R}^2} h_\Delta^{(2)}(u):x^{\otimes 2}:(u)du \\ + \int_{\mathbb{R}^4} h_\Delta^{(4)}(u):x^{\otimes 4}:(u)du\end{aligned} \tag{3.24}$$

with

$$h_\Delta^{(o)} = (g_\Delta^{(o)})^2 + \int_{\mathbb{R}^2} g_\Delta^{(2)}(u_1,u_2)g_\Delta^{(2)}(u_2,u_1)du_1du_2 \tag{3.25}$$

$$h_\Delta^{(2)} = g_\Delta^{(2)}(u_1,\cdot)*g_\Delta^{(2)}(u_2,\cdot) + 2g_\Delta^{(2)}(u_1,\cdot)*g_\Delta^{(2)}(\cdot,u_2) + 2g_\Delta^{(2)}(u_1,u_2)g_\Delta^{(o)} \tag{3.26}$$

$$h_\Delta^{(4)} = (g_\Delta^{(2)})^{\otimes 2}(u) \tag{3.27}$$

and we used the notation

$$g_\Delta^{(2)}(u_1,\cdot)*g_\Delta^{(2)}(u_2,\cdot) = \int g_\Delta^{(2)}(u_1,s)g_\Delta^{(2)}(u_2,s)ds \tag{3.28}$$

etc.

Next we argue that the last three terms on the right hand side of (3.24) vanish weakly in (L^2) as $o(\varepsilon)$ (uniformly in t). This is obvious for $h_\Delta^{(o)}$ since our assumptions on φ imply that $s \mapsto f(s;s+)$ and $s \mapsto f(s;\cdot) \in L^2(\mathbb{R})$ are continuous. Consider one of the terms coming from $h_\Delta^{(2)}$:

$$\psi_\Delta^{(2)}(x) := \int_{\mathbb{R}^2} g_\Delta^{(2)}(u_1,\cdot)*g_\Delta^{(2)}(\cdot,u_2):x(u_1)x(u_2):du_1du_2 \tag{3.29}$$

It takes a moment's thought to see that because $\varepsilon^{-1}\psi_\Delta^{(2)}$ in (L^2)-bounded $\varepsilon^{-1}\psi_\Delta^{(2)} \to 0$ weakly iff $\varepsilon^{-1}S\psi_\Delta^{(2)}(\xi) \to 0$ for every $\xi \in S(\mathbb{R})$. But

$$S\psi_\Delta^{(2)}(\xi) = \int_\Delta \left(\int_\Delta f(u_1;s)\xi(u_1)du_1\right)\left(\int f(s;u_2)\xi(u_2)du_2\right)ds \tag{3.30}$$

which one easily estimates (using Schwarz' inequality) as follows

$$|S\psi_\Delta^{(2)}(\xi)| \le \|\xi\|_2^2\left(\int_\Delta \|1_\Delta(\cdot)f(s;\cdot)\|_2^2 ds\right)^{1/2}\left(\int_\Delta \|f(s;\cdot)\|_2^2 ds\right)^{1/2} = \|\xi\|_2^2\, o(\varepsilon) \tag{3.31}$$

due to our assumptions on f. A compactness argument can be used to show that the estimate is uniform in t. The other terms coming from $h_\Delta^{(2)}$ and the one with $h_\Delta^{(4)}$ in (3.24) are treated similarly. In order to prove (3.20) it therefore remains to show that

$$\varepsilon^{-1}\int_\Delta \varphi(u)^2 du \rightarrow \varphi(t)^2 \tag{3.32}$$

weakly and uniformly in t. But since $s \rightarrow \varphi(s)^2 \in (L^2)$ is continuous on $[0,1]$ this is trivial.

Finally we can prove (3.15) in our simplified case. It suffices to show that for every $\psi \in (L^2)$

$$\begin{aligned}(F(X(t))-F(X(s)) - \int_s^t F'(X(u))dX(u+) + \\ + \frac{1}{2}\int_s^t F''(X(u))\varphi(u)^2 du, \psi)\end{aligned} \tag{3.33}$$

is zero. (3.33) is equal to (using a partition of $[s,t]$ as in (3.17))

$$\begin{aligned}(\sum_k F'(X(u_k))\Delta_k X - \int_s^t F'(X(u))dX(u+), \psi) + \\ + \frac{1}{2}(\sum_k F''(X(u_k))(\Delta_k X)^2 - \int_s^t F''(X(u))\varphi(u)^2 du, \psi)\end{aligned} \tag{3.34}$$

(neglecting the third order term). The first term of (3.34) vanishes in the limit that the mesh of the partition tends to zero (cf. (3.14)). The second one we write in the following way

$$\begin{aligned}\frac{1}{2}\{\sum_2 (F''(X(u_k))\varphi(u_k)^2, \psi)|\Delta_k| - \int_s^t (F''(X(u))\varphi(u)^2, \psi)du\} + \\ + \frac{1}{2}\sum_k |\Delta_k|\{(F''(X(u_k))(|\Delta_k|^{-1}(\Delta_k X)^2 - \varphi(u_k)^2), \psi)\}\end{aligned} \tag{3.35}$$

The first of the last two terms vanishes in the limit $\sup_k |\Delta_k| \rightarrow 0$ because by assumption $u \rightarrow (F''(X(u))\varphi(u)^2, \psi)$ is continuous on $[0,1]$ for every $\psi \in (L^2)$ and therefore the sum is the Riemann approximation to the integral. Concerning the last term in (3.35) we use the fact that $|\Delta|^{-1}(\Delta X)^2 \rightarrow \varphi(t)^2$, $\Delta \ni t$, $|\Delta| \rightarrow 0$, weakly and uniformly in t.

This together with a compactness argument implies that the scalar product in this expression vanishes uniformly in k. Thus we have shown that (3.33) is zero for every $\psi \in (L^2)$ and so (3.15) holds in (L^2).

REFERENCES

[Ad 75] Adams, R.A.: Sobolev spaces; New York: Academic Press (1975)

[AP 88] Asch, J., Potthoff, J.: 'Itô's lemma without non-anticipatory conditions'; BiBoS preprint (1988)

[Ch 86] Chow, P.L.: 'Generalized solution of some parabolic equations with a random drift'; preprint (1986), to appear in J. Appl. Math. Optimization

[Hi 70] Hida, T.: Stationary stochastic processes; Princeton: Princeton University Press (1970)

[Hi 75] Hida, T.: Analysis of Brownian functionals; Carleton Mathematical Lecture Notes, 13 (1975)

[Hi 80] Hida, T.: Brownian motion; Berlin, Heidelberg, New York: Springer (1980)

[HPS 88a] Hida, T., Potthoff, J., Streit, L.: 'Dirichlet forms and white noise analysis'; Commun. Math. Phys., 116 (1988) 235-245

[HPS 88b] Hida, T., Potthoff, J., Streit, L.: 'White noise analysis and applications'; Mathematics + Physics, 3; ed. by L. Streit, Singapore: World Scientific, to appear

[HS 77] Hida, T., Streit, L.: 'On quantum theory in terms of white noise'; Nagoya Math. J., 68 (1977) 21-34

[Kub 83] Kubo, I.: 'Itô formula for generalized Brownian functionals'; in: Theory and application of random fields; ed. by G. Kallianpur, Berlin, Heidelberg, New York: Springer (1983)

[KT 80] Kubo, I., Takenaka, S.: 'Calculus on Gaussian white noise, I-IV'; Proc. Japan Acad., 56 (1980) 376-380, 411-416, 57 (1981) 433-437, 58 (1982) 186-189

[Ku 83] Kuo, H.-H.: 'Brownian functionals and applications'; Acta Appl. Math., 1 (1983) 175-188

[KR 88] Kuo, H.-H., Russek, A.: 'White noise approach to stochastic integration'; J. Multivariate Anal., 24 (1988) 218-236

[KS 87] Kuo, H.-H., Shieh, N.-R.: 'A generalized Itô's formula for multi-dimensional Brownian motions and its application'; Chinese J. Math., 15 (1987) 163-174

[Ne 73] Nelson, E.: 'Probability theory and Euclidean quantum field theory'; in: Constructive quantum field theory; ed. by G. Velo and A. Wightman; Berlin, Heidelberg, New York: Springer (1973)

[NP 88] Nualart, D., Pardoux, E.: 'Stochastic calculus with anticipating integrands'; preprint (1988)

[PP 87] Pardoux, E., Protter, P.: 'A two-sided stochastic integral and its calculus'; Probab. Th. Rel. Fields, 76 (1987) 15-49

[Po 87] Potthoff, J.: 'White noise approach to Malliavin calculus'; J. Funct. Anal., 71 (1987) 207-217

[Po 88] Potthoff, J.: 'On Meyer's equivalence'; to appear in Nagoya Math. J., (1988)

[Si 74] Simon, B.: The $P(\varphi)_2$ Euclidean (quantum) field theory; Princeton: Princeton University Press (1974)

[Str 88] Streit, L.: contribution to this volume

[Tak 83] Takenaka, S: 'Invitation to white noise calculus'; in: Theory and application of random fields; ed. by G. Kallianpur, Berlin, Heidelberg, New York: Springer (1983)

[Yo 87] Yokoi, Y.: 'Positive generalized functionals', Preprint (1987)

QUELQUES REMARQUES SUR LE CUT-LOCUS SOUS-RIEMANNIEN

G. BEN AROUS
Département de Mathématiques
École Normale Supérieure
45 rue d'Ulm
75230 PARIS CEDEX 05

R. LÉANDRE
Département de Mathématiques
Faculté des Sciences
16 route de Gray
25030 BESANÇON CEDEX

INTRODUCTION

Nous voulons ici illustrer le théorème (de [B-A.1] et [L.1]) sur le comportement asymptotique en temps petit du noyau de la chaleur $p_t(x,y)$ associé à un opérateur

$$L = 1/2 \, \Sigma_{1 \le i \le m} X^2_i + X_0$$

(où les X_i sont des champs de vecteurs C^∞ sur $\mathbf{R}^d$ vérifiant l'hypothèse de Hörmander forte : dim Lie $(X_1 \ldots X_m)(x) = d$, $\forall x \in \mathbf{R}^d$).

Ce résultat affirme, comme le conjecturait Bismut [B], que, pour des points (x,y) hors du cut-locus sous-riemannien (dont la définition est rappelée plus bas), le noyau $p_t(x,y)$ se comporte en temps petit comme si l'opérateur L était elliptique.

Nous allons, dans un cadre assez simple, donner une description explicite du cut-locus.Cette description nous permettra de montrer que le résultat de [B-A.1] et [L.1] n'est pas du tout de nature "presque elliptique" . Elle nous permettra surtout de montrer qu'il est impossible de définir un rayon d'injectivité sous-riemannien. Plus précisément, si deux points x,y sont hors du cut-locus, il se peut que, pour un point z situé entre x et y sur la géodésique joignant x à y , alors x et z soient conjugués (contrairement à ce qui se passe dans le cas riemannien).Ceci montre la difficulté du passage d'estimation (fines) locales à des estimations (fines) globales pour le noyau de la chaleur hypoelliptique et explique la nécessité des méthodes directement globales utilisées par [B-A.1] et [L.1].

Pour un rappel plus complet sur les résultats relatifs au noyau de la chaleur hypoelliptique, nous renvoyons aux articles de revue [B-A.2] et [L.2].

S. Albeverio et al. (eds.), Stochastics, Algebra and Analysis in Classical and Quantum Dynamics, 15–22.

1. RAPPELS ET NOTATIONS

L'essentiel des rappels faits ici est tiré de [B].
Considérons sur $\mathbf{R}^d$ m champs de vecteurs $X_1 \ldots X_m$ de classe C^∞ vérifiant l'hypothèse de Hörmander forte :

(1.1) $\quad \forall x \in \mathbf{R}^d \quad \dim \mathrm{Lie}(X_1 \ldots X_m)(x) = d .$

Ces m champs définissent sur $\mathbf{R}^d$ une métrique sous-riemannienne de la façon suivante :
Soit H^1 l'ensemble des fonctions absolument continues sur $[0,1]$, à valeurs dans $\mathbf{R}^m$, nulles en 0, dont la dérivée est de carré intégrable. Pour $h \in H^1$, on notera h' la dérivée de h et $|h|_1 = (\int h'^2_s \, ds)^{1/2}$ sa norme.
Pour $h \in H^1$, considérons le flot horizontal φ^h défini sur $\mathbf{R}^d$ par l'équation différentielle

$$(1.2) \qquad \begin{aligned} d\varphi_t(x)/dt &= \Sigma_{1 \le i \le m} X_i(\varphi_t(x))\, h'^i_t \\ \varphi_0(x) &= x \end{aligned}$$

et notons ϕ_x l'application de H^1 dans $\mathbf{R}^d$, qui à $h \in H^1$, associe $\varphi^h{}_1(x)$.

Pour $(x,y) \in \mathbf{R}^d \times \mathbf{R}^d$, on note $K(x,y) = \phi_x^{-1}(\{y\})$ l'ensemble des contrôles $h \in H^1$ qui définissent des courbes horizontales joignant x à y.
Si l'on pose :

$$(1.3) \qquad d(x,y) = \inf(|h|_1 , h \in K(x,y)),$$

on sait que $d(x,y)$ est une distance continue sur $\mathbf{R}^d$ et que l'infimum dans (1.3) est toujours atteint.Pour $h \in K(x,y)$ tel que $|h|_1 = d(x,y)$, on dira naturellement que la courbe horizontale $\varphi^h(x)$ est une géodésique (minimisante) joignant x à y.

Une question naturelle est de décrire précisément ces géodésiques et en particulier de relier la formulation lagrangienne du minimum (1.3) à une formulation hamiltonienne. Plus précisément, soit $H(x,p)$ le hamiltonien défini sur $T^*\mathbf{R}^d$ par

$$(1.4) \qquad H(x,p) = 1/2 \; \Sigma_{1 \le i \le m} \langle X_i(x),p\rangle^2$$

et $\psi_t(x,p) = (x_t(x,p),p_t(x,p))$ le flot hamiltonien (ou bicaractéristique) associé :

$$(1.5) \qquad \begin{aligned} dx_t/dt &= \partial_p H(x_t,p_t) \\ dp_t/dt &= -\partial_x H(x_t,p_t) \end{aligned} \qquad \text{avec} \qquad \begin{aligned} x_0 &= x \\ p_0 &= p . \end{aligned}$$

Notons π la projection de $T^*\mathbf{R}^d$ sur $\mathbf{R}^d$; il est clair que pour que $(x,p) \in T^*\mathbf{R}^d$ la projection $\pi\psi_t(x,p)$ d'une bicaractéristique est une courbe horizontale au sens de (1.2). La réciproque n'est pas évidente. On doit à Bismut [B] la réciproque partielle suivante :

THÉORÈME (1.6).- *Si l'application* ϕ_x *est une submersion en* $h \in H^1$ *, i.e.* :

(1.7) $$d\phi_x(h)(H^1) = \mathbf{R}^d ,$$

alors la courbe horizontale $\varphi^h(x)$ *est projection d'une bicaractéristique unique.*

Il est facile de vérifier que la condition de submersion (1.7) est équivalente à la non-dégénérescence de la forme quadratique définie sur $\mathbf{R}^d$ par :

(1.8) $$<C^{h,x}p,p> = \Sigma_{1\le i\le m} \int_{0\le s\le 1} <(\varphi^h{}_s)^{*-1} X_i(x),p>^2 ds.$$

Cette forme quadratique (ou plutot la matrice $C^{h,x}$) sera appelée dans la suite matrice de Malliavin déterministe de h .
On définit alors la notion de cut-locus sous-Riemannien :

DÉFINITION (1.9).- *On dira que* (x,y) *n'appartient pas au cut-locus si et seulement si* :
a)*Il existe une unique géodésique minimisante joignant* x *à* y :
i.e. $\exists !\ h \in K(x,y),\ \ |h|_1 = d(x,y)$.
b)*Cette géodésique est la projection d'une unique bicaractéristique* :
i.e. $\exists !\ p_0 \in T_x{}^* \mathbf{R}^d ,\ \ \pi\psi_t(x,p_0) = \varphi^h{}_t(x) \quad \forall t \in [0,1]$.
c) *Les points* x *et* y *sont non conjugués le long de cette géodésique* :
i.e. $\partial_p\, \pi\psi_1(x,p_0)$ *est inversible.*

***Remarque* (1.10).-** Dans cette définition, on peut remplacer b) par
b') ϕ_x est une submersion en h
ou, comme on l'a vu, par : $C^{x,h}$ est non dégénéré.En effet, on a vu que b') $\Rightarrow$ b).
Réciproquement, sous l'hypothèse b), on a : $\partial_p\, (\pi\psi(x,p_0))(T_x{}^* \mathbf{R}^d) \subset d\, \phi_x(h)(H')$ et donc sous l'hypothèse c), b') est vérifié.

L'intérêt de la notion de cut-locus est contenu dans les remarques suivantes :

1)Dans le cas Riemannien (ou elliptique), c'est-à-dire dans le cas où l'espace vectoriel engendré par les champs $X_1 \ldots X_m$ est de dimension d en tout point, la métrique définie par (1.3) coïncide avec la métrique riemannienne associée à la structure riemannienne : $g_{ij}(x) = \Sigma_{1\le k\le m} X_k{}^i(x)\, X_k{}^j(x)$, et la définition (1.9) du cut-locus coïncide avec la définition usuelle, car alors l'hypothèse b') est toujours satisfaite.

2)Comme dans le cas Riemannien, le cut-locus est fermé; sur son complémentaire $d^2(x,y)$ est C^∞ en (x,y) et la géodésique qui joint x à y varie de façon C^∞ avec (x,y) .

3)Enfin et surtout, sur le complémentaire du cut-locus, le comportement asymptotique, lorsque t tend vers zéro, du noyau de la chaleur $p_t(x,y)$ associé à l'opérateur $L = 1/2\, \Sigma_{1\le i\le m} X_i{}^2 + X_0$ (où X_0 est un autre champ de vecteur C^∞ sur $\mathbf{R}^d$) est identique au comportement asymptotique du noyau de la chaleur Riemannien.
Précisément, on a, pour tout n > 0 :

$$p_t(x,y) = 1/t^{d/2} \exp(- d^2(x,y)/2t) \ (\Sigma_{0\le k\le n} c_k(x,y)t^k + 0(t^{n+1}))$$

où les c_k sont C^∞ sur un voisinage de (x,y) et où $c_0(x,y) > 0$ (voir [B-A.1] et [L.1]).

2. CARACTÉRISATION DU CUT-LOCUS DANS UNE SITUATION SIMPLE

Pour un point x de $\mathbf{R}^d$, notons x_1 sa première coordonnée et $x' = (x_2 \ldots x_d)$. Considérons des champs de vecteurs $X_1 \ldots X_m$ sur $\mathbf{R}^d$ tels que

(2.1) $\quad X_1 = \partial/\partial x_1$ et X_k est tangent à $\{X_1 = 0\} = \mathbf{R}^{d-1}$ pour $2 \leq k \leq m$.

Notons $X_k = \Sigma_{2\leq j\leq d} X^j{}_k(x_1,x')\partial/\partial x_j$ pour $2 \leq k \leq m$.

Dans cette situation, on a :

***Lemme* (2.2).-**
a) $\quad d(x,y) \geq |y_1 - x_1|$.

b) *Si* $x' = y'$ *, l'unique géodésique joignant* x *à* y *est le segment de droite, et*

$$d(x,y) = |x_1 - y_1| = \|y - x\|.$$

Preuve.- **a)** Soit $h \in H^1$ et $\varphi_t{}^h(x)$ la courbe horizontale associée. Il est clair que la première composante de $\varphi_t{}^h(x)$ est donnée par : $[\varphi_t{}^h(x)]_1 = x_1 + h^1(t)$,

ainsi si $h \in K_{x,y}$ on a: $h^1(t) = y_1 - x_1$

et donc, pour $h \in K_{x,y}$: $|h|_1 = (\int h'^1(s)^2\, ds)^{1/2} \geq |h^1(1) - h^1(0)| = |y_1 - x_1|$,
d'où $d(x,y) \geq |y_1 - x_1|$.

b) Si $y' = x'$, il est clair que si l'on pose

$$h^1(t) = x_1 + t(y_1 - x_1)$$
$$h^j(t) \equiv 0 \quad \text{pour } 2 \leq j \leq m$$

on a $h \in K_{x,y}$ et $|h|_1 = |y_1 - x_1|$ et donc par le a) : $d(x,y) = |y_1 - x_1|$

φ^h est l'unique géodésique joignant x à y (φ^h est, bien sur, le segment de droite joignant x à y, parcouru en temps 1).

On peut alors, pour un tel couple, caractériser très simplement le fait qu'il est dans le cut-locus. Posons $y_t = (x_1 + t(y_1 - x_1),x')$ et $V_t(x,y) = \mathrm{Vect}(X_k(y_t),\ 2 \leq k \leq m)$

$V_t(x,y)$ est l'espace vectoriel engendré au point $\varphi_t{}^h(x)$ par les champs $X_2 \ldots X_m$ transverses à la géodésique.

THÉORÈME (2.4).- *Soient* $x,y \in \mathbf{R}^d$ *tels que* $x' = y'$; *les propositions suivantes sont équivalentes* :

a) (x,y) *n'appartient pas au cut-locus* ;
b) $\dim \oplus_{t\in[0,1]} V_t(x,y) = d - 1$

Remarque : La condition b), qui exprime que les champs transverses à la géodésique doivent remplir tout l'espace transverse, est intéressante car elle ne fait pas intervenir de crochets des champs X_i . Remarquons aussi qu'on lit facilement sur la condition b) la propriété suivante :

COROLLAIRE 2.5.- *Soient* x,y *deux points de* $\mathbf{R}^d$ *tels que* $x' = y'$ *et tels que* (x,y) *soit hors du cut-locus. Alors si, pour tout* $t \in \mathbf{R}^+$, *on note* $y_t = (x_1 + t(y_1 - x_1),x')$, *il existe un* $T < 1$ *tel que, pour tout* $t \in]T,\infty[$, *le couple* (x,y_t) *est hors du cut-locus.*

***Preuve du corollaire* 2.5.-** Le cut-locus étant fermé, il est clair qu'il existe $T < 1$ tel que, pour $t \in]T,1[$, (x,y_t) est hors du cut-locus.
D'autre part, la dimension d(t) de $\oplus_{s\in [0,1]} V_s(x,y_t)$ étant croissante, il est clair que pour tout $t \geq 1$ elle vaut $d - 1$. D'où le résultat.

On a enfin, par [B-A.1] et [L.1] :

COROLLAIRE 2.6.-
Soient $x,y \in \mathbf{R}^d$ *tels que* $x' = y'$ *et tels que* : $\dim \oplus_{t\in [0,1]} V_t(x,y) = d - 1$;
alors on a le développement asymptotique du noyau de la chaleur, pour tout n :

$$p_t(x,y) = 1/t^{d/2} \exp(-(d^2(x,y)/2t)(\Sigma_{0\leq k\leq n} c_k(x,y)\, t^k + 0(t^{n+1}))$$

avec $c_0(x,y) > 0$; *et ce développement est valide (et uniforme)sur un voisinage de* (x,y) .

***Preuve du théorème* (2.4).-** Considérons le flot hamiltonien $\psi_t(x,p)$ défini en (1.5).
Pour $p \in T_x{}^*\mathbf{R}^d$, on notera encore $p = (p_1,p')$.
On a ici :

$$H(x,p) = 1/2\,(p_1{}^2 + \Sigma_{2\leq i\leq m} <X_i(x_1,x'),p'>^2) .$$

L'équation (1.5) s'écrit alors :

$$dx_1(t)/dt = p_1 \;\; ; \;\; dx_l(t)/dt = \Sigma_{2\leq i\leq m} (\Sigma_{2\leq j\leq m} X_i{}^j(x(t))p_j(t))\, X_i{}^l(x(t)) \;\text{ pour } l \geq 2$$

et $\forall k \in \{1...s\}$:

$$dp_k(t)/dt = -\Sigma_{2\leq i\leq m} (\Sigma_{2\leq j\leq m} X_i{}^j(x(t))p_j(t))\, (\Sigma_{2\leq j'\leq m} (\partial/\partial x_k\, X_i{}^{j'}(x(t))p_{j'}(t))$$

avec $x(0) = x$ et $p(0) = p$.

Choisissons $p(0) = p$ orthogonal à $\mathbf{R}^{d-1}$; i.e. $p' = 0$.
On a alors $p'(t) = 0$ et $p_1(t) = p_1$ pour tout $t \in [0,1]$
et donc $x_1(t) = x_1 + p_1 t$, $x'(t) = x'$ pour tout $t \in [0,1]$.
En prenant $p_1 = y_1 - x_1$, on vérifie donc que la géodésique joignant x à y est bien la projection d'une unique bicaractéristique (si $x_1 \neq y_1$).
Pour décider si (x,y) est dans le cut-locus ou pas, il s'agit d'étudier la validité du c) de la définition (1.9), et donc de calculer le jacobien $\partial_p\, \pi\psi_1(x,p)$.
Pour cela on a, du fait que $p' = 0$:

$$(d/dt)\,(\partial x_1(t)/\partial p_1) = 1 \;\; ; \;\; (d/dt)\,(\partial x_1(t)/\partial p_l) = 0 \;\text{ pour } l \geq 2$$

et

$$(d/dt)\,(\partial x_j(t)/\partial p_l) = \Sigma_{2\leq i\leq m} X_i{}^l(x(t))\, X_i{}^j(x(t)) \;\text{ pour } j \geq 2 \text{ et } 1 \leq l \leq d .$$

On obtient ainsi que la matrice jacobienne $\partial_p \pi\psi_1(x,p)$ est de la forme:

$$\begin{matrix} 1 & * \cdots\cdots\cdots * \\ 0 & \\ \cdot & \\ \cdot & A \\ \cdot & \\ 0 & \end{matrix}$$

où la matrice $(d-1)x(d-1)$ A est donnée par : $A_{kj} = \int_{0\le t\le 1} A_{kj}(t)dt$ avec

$$A_{kj}(t) = \Sigma_{2\le i\le m} X_i^k(y_t)\, X_i^j(y_t) .$$

A est une matrice symétrique, positive.
Elle est définie positive si et seulement si l'intersection des noyaux des formes quadratiques définies par les matrices $A(t)$ est réduite à 0 . Ce qui est clairement équivalent à la condition b).

3. EXEMPLES

Nous allons donner ici deux exemples extrêmes d'application du théorème (2.4) qui illustrent le fait que le résultat de développement asymptotique du noyau de la chaleur hors du cut- locus est lié à un phénomène plus subtil que l'ellipticité. Nous allons donner un premier exemple (déjà traité dans [B-A.2]) où ce développement asymptotique est valide alors que l'opérateur est très loin d'être elliptique (précisément, il faut un nombre arbitrairement grand de commutateurs des X_i pour engendrer l'espace tangent) puis un autre où, au contraire, le développement asymptotique n'est pas valide alors que l'opérateur est elliptique presque partout et qu'il suffit d'un commutateur de longueur 2 pour engendrer l'espace tangent.

__Exemple 1__ :
Soient les deux champs sur $\mathbf{R}^d$:

$$X_1 = \partial/\partial x_1 \quad \text{et} \quad X_2 = \Sigma_{2\le i\le d}\, x_1^{k_i}\, \partial/\partial x_i$$

où les k_i sont des entiers tels que $0 < k_2 < \ldots < k_d$.
Alors pour deux points *distincts* x,y tels que $x' = y'$ la condition b) est toujours vérifiée.
En effet, l'espace vectoriel $V_t(x,y)$ est ici la droite vectorielle engendrée par

$$X_2(t) = X_2(x_1 + t(y_1 - x_1), x') .$$

Et il est clair que si $0 < t_1 < \ldots < t_{d-1} < 1$, les vecteurs $X_2(t_1) \ldots X_2(t_{d-1})$ sont indépendants. Ainsi $\dim \oplus_{t\in[0,1]} V_t(x,y) = d - 1$.

Ainsi dans cette situation arbitrairement dégénérée, le noyau de la chaleur associé à $L = 1/2\,(X_1^2 + X_2^2)$ se comporte en temps petit, pour des points x,y distincts tels que $x' = y'$ (et pour les points voisins, puisque le cut-locus est fermé), comme si l'opérateur était elliptique.
Il est possible de retrouver ce résultat directement, dans ce cas, à partir du résultat de grandes déviations pour le pont brownien (voir [B-A.2]).

Exemple 2 :
Considérons sur $\mathbf{R}^3$ les trois champs :

$$X_1 = \partial/\partial x_1$$
$$X_2 = \partial/\partial x_2$$
$$X_3 = x_2\, \partial/\partial x_3 \qquad \text{et l'opérateur } L = 1/2 \sum_{1\leq i\leq 3} X_i^2 .$$

La situation est ici très différente. L est elliptique partout sauf sur le plan $\{x_2 = 0\}$, et les champs X_1, X_2, $[X_2,X_3]$ engendrent partout l'espace tangent. La situation semble donc beaucoup plus proche d'une situation elliptique ; nous allons pourtant vérifier que le développement asymptotique "de type elliptique" donné par le corollaire (2.6) n'est pas toujours vrai.
Soient deux points x,y tels que $x' = y'$; alors (x,y) est dans le cut-locus si et seulement si $x_2 = y_2 = 0$.

En effet, l'espace $V_t(x,y)$ est alors engendré par les vecteurs $X_2(y_t)$ et $X_3(y_t)$. Cet espace est indépendant de t. C'est une droite vectorielle si $x_2 = 0$, un plan si $x_2 \neq 0$; d'où le résultat.
En fait, il est possible de donner ici l'équivalent précis du noyau de la chaleur $p_t(x,y)$ associé à L pour des points x,y tels que $x' = y'$ et $x_2 = 0$.
Considérons en effet l'opérateur de Grushin sur $\mathbf{R}^2$:

$$L = (\partial/\partial x_2)^2 + x_2^2\, (\partial/\partial x_3)^2$$

et $q_t(x',y')$ son noyau de la chaleur.

Si $x_2 = 0$, on a (*cf.* [B-A.3]) $q_t(x',x') = \text{constante}/t^{3/2}$

et si $x_2 \neq 0$, on a : $\qquad q_t(x',x') \sim c(x')/t \qquad$ avec $c(x') > 0$.

Or on a, de façon évidente:

$$p_t(x,y) = q_t(x',y') \exp(-(x_1 - y_1)^2/2t) / (2\pi t)^{1/2} .$$

D'où, si $x' = y'$, on a, si $x_2 \neq 0$

$$p_t(x,y) \sim c(x')/\sqrt{2\pi}\, t^{3/2} \exp(-(x_1 - y_1)^2/2t)$$

ce qui est le résultat du corollaire (2.6) (et du lemme 2.2. b)).
Par contre, si $x_2 = 0$, on obtient :

$$p_t(x,y) = \text{constante}/t^2 \exp(-(x_1 - y_1)^2/2t)$$

ce qui montre alors que le résultat du corollaire (2.6) est bien faux (car ici (x,y) est dans le cut- locus).

4. LE PROBLÈME DU RAYON D'INJECTIVITÉ

Dans le cas d'une métrique riemannienne, on sait que si x et y sont hors du cut-locus, et si l'on note φ_t la géodésique minimisante joignant x à y en temps 1, alors pour tout couple (s,t) tels que $0 \le s < t \le 1$ (φ_s,φ_t) est hors du cut-locus ; i.e. si deux points sont non conjugués, alors les points intermédiaires sur la géodésique qui les joint sont aussi non conjugués.
On va vérifier qu'il n'en est rien dans le cas sous-riemannien, en utilisant la caractérisation précédente du cut-locus dans le cas simple traité ici.
Précisément, il s'agit de construire un exemple de points x,y (avec x' = y') hors du cut-locus tel que le T introduit au corollaire (2.5) est strictement positif (l'exemple 1 donne un exemple de situation où T = 0).
Considérons pour cela une fonction de troncature ρ, C^∞ à support dans $]-\infty,1/2]$, à valeurs dans $[0,1]$, strictement positive sur $]-\infty,1/2[$ et égale à 1 sur $]-\infty,1/4[$.
Soient alors les champs sur $\mathbf{R}^3$ donnés par :

$$X_1 = \partial/\partial x_1$$
$$X_2 = [\rho(x_1) + (1-\rho(x_1))x_1]\ \partial/\partial x_2$$
$$X_3 = [\rho(x_1)x_2 + (1-\rho(x_1))x_1^2]\ \partial/\partial x_3$$

et considérons les points $x = (0,0,0)$ et $y = (1,0,0)$.
Pour $x_1 < 1/4$, les champs X_1, X_2, X_3 coïncident avec ceux de l'exemple 2.
Pour $x_1 > 1/2$, ils coïncident avec ceux de l'exemple 1 (dans un cas particulier).

Si l'on note $y_t = (t,0,0)$ un point de la géodésique joignant x à y (avec $t \in [0,1]$), on a, de façon évidente :
si $t < 1/4$: (x,y_t) est dans le cut-locus,
si $t > 1/2$: (x,y_t) n'est pas dans le cut-locus.
Ce qui donne bien l'exemple cherché.

BIBLIOGRAPHIE:

[B-A.1] G.BEN AROUS - *Développement asymptotique du noyau de la chaleur hors du cut-locus*, à paraître aux Annales de l'E.N.S.
[B-A.2] G.BEN AROUS - *Noyau de la chaleur hypoelliptique et géométrie sous-Riemannienne,* a paraître aux Actes du Colloque franco-japonais,Paris 1987.
[B-A.3] G.BEN AROUS - *Développement asymptotique du noyau de la chaleur sur la diagonale*, à paraître aux Annales de l'Institut Fourier.
[B] J.-M. BISMUT - *Large deviations and the Malliavin calculus*, Progress in Math. n° 45, Basel-Boston-Stuttgart Birkhaüser (1984).
[L.1] R. LÉANDRE - *Intégration dans la fibre associée à une diffusion dégénérée*,à paraître in Prob. Theo. related fields.
[L.1] R. LÉANDRE - *Applications quantitatives et géométriques du calcul de Malliavin*, à paraître aux Actes du Colloque franco-japonais,Paris 1987. Version anglaise à paraître aux Proceedings du Colloque "Geometry of Random Motion".

SIMULATION ON RANDOM GRAPHS OF THE EPIDEMIC DYNAMICS OF SEXUALLY TRANSMITTED DISEASES - A NEW MODEL FOR THE EPIDEMIOLOGY OF AIDS *)

Georg F. Bolz
Fakultät für Physik and BiBoS
Universität Bielefeld
D-4800 Bielefeld 1

ABSTRACT. A new model to describe the epidemic dynamics of sexually transmitted diseases is described; its main ingredients are: a random graph, encoding the sexual contact structure of a society, and a discrete time stochastic process evolving on it, representing the spread of infection. The appropriateness, in the case of AIDS, of this approach will be motivated and first results from computer simulations for a set of simplistic test assumptions will be presented.

1. INTRODUCTION

In this talk I'm going to present a novel approach to modelling the epidemic dynamics of sexually transmitted diseases, initiated by and aimed primarily at an attempt to improve our understanding of the peculiar features of the AIDS (acquired immunodeficiency syndrom)-epidemic or, better to say, of the underlying epidemic of HIV (human immunodeficiency virus)-infection. This approach has been developed in collaboration with Philippe Blanchard and Tyll Krüger and first presented in [BBK]. This presentation will be oriented towards the simulational aspects of our modelling approach.

Since, here, I am addressing an audience of mathematical physicists, I will start with an overview over the epidemic situation; clearly, I cannot go into a detailed discussion of the present data nor the history of AIDS; rather, those aspects of the HIV-infection and of the conditions of its spread which constitute the particular features of the AIDS-epidemic will be noted. After calling to mind standard epidemiological models by means of a few prototypes, I will present the ideas behind our approach heuristically, while the following paragraph will contain a description in mathematical terms. Next, I will dare to connect, through a few cautious remarks, with random graph theory and percolation theory. The central part of this talk is then the

*) The work underlying this report has been supported by Bundesminister für Forschung und Technologie under contract number II-043-88. The responsability for the content lies with the author.

S. Albeverio et al. (eds.), Stochastics, Algebra and Analysis in Classical and Quantum Dynamics, 23–51.

description of the computer simulations and the presentation of results from a first toy model. I will end with a few remarks about future work and lines of thought.

2. HIGHLIGHTING THE EPIDEMIC SITUATION OF THE HIV-INFECTION

AIDS, the acquired immunodeficiency syndrom, has been described as such first in 1981 [Si], while early cases can be traced back to the mid-70, probably even further. Revealing the role of HIV, the human immunodeficiency virus, as the causative agent of the deterioration of the protective capabilities of the immune system, letting a person succumb to a plethora of opportunistic infections, could be accomplished very rapidely due to the fact that AIDS appeared before our eyes just at the time when we had acquired the tools to identify a retrovirus whose life is so intricately linked with the very working of our immune system. Thus the main traits of the HIV-infection are now clearly understood. However a lot of important and epidemiologically relevant details remain still unknown. Moreover, when we try to describe in a quantitative way the scene which the process of transmission evolves on, we are facing an astonishing and maybe embarrassing ignorance about patterns of sexual behaviour in our societies. Thirdly, a lot of uncertainty resides in the available data about the extent to which certain subpopulations of the society are affected by HIV-infection and, even more so, about the actual patterns of transmission along the known channels.

While these unknowns put a heavy burden on modelers and make true forecasting impossible at this time, they do not diminish the crucial role epidemiological models have to play in understanding the dynamics of the AIDS-epidemic nor do they constitute a reason not to look for a modelling framework which is detailed enough to be able to include all the relevant features of the epidemic dynamics of HIV-infection. It is the central hypothesis of our approach that the structure of the mesh of sexual contacts inside a population does have a determining effect on the epidemic dynamics of a sexually transmitted disease.

2.1. A Look at the Last Reported Figures

Clearly, I cannot here review what is known and what is not about the epidemic situation around the world nor is it needed for the purpose of this talk. Thus let us be content with having a short exemplary look at the figures for the Federal Republic (table 1) and at the global figures (table 2) as they are known at the time of this talk.

	Registrierte AIDS-Fälle im Zeitraum vom 1.1.82-29.1.88			**Neuregistrierte AIDS-Fälle im Zeitraum vom**				
				1.1.82-31.12.84	1.1.-31.12.85	1.1.-31.12.86	1.1.-31.12.87	
	gesamt	Frauen	verstorben	gesamt	gesamt	gesamt	gesamt	verstorben
Baden-Württemberg	111	21	55	8	25	24	52	20
Bayern (*München*)	289 (*220*)	9 (*4*)	137 (*105*)	23 (*19*)	40 (*35*)	70 (*58*)	136 (*92*)	48 (*31*)
Berlin	367	15	142	25	36	80	207	56
Bremen	34	1	11	3	6	2	22	8
Hamburg	150	4	63	13	33	32	67	12
Hessen (*Frankfurt*)	265 (*180*)	18 (*11*)	138 (*91*)	25 (*19*)	33 (*22*)	69 (*45*)	121 (*84*)	29 (*17*)
Niedersachsen	88	9	49	0	11	27	45	17
Nordrhein-Westfalen	360	24	146	22	37	89	194	54
Rheinland-Pfalz	54	5	17	1	3	16	32	7
Saarland	18	2	7	0	2	4	10	4
Schleswig-Holstein	24	1	4	0	2	4	18	2
Gesamt	1760	109	769	120	228	417	904	257

a) b)

Risikogruppe	Fallzahl männl.	Fallzahl weibl.	% gesamt
1. Homo- oder Bisexuelle Männer	1296	-	73,6
2. Fixer	92	59	8,6
2. a) Risiken 1.+2.	19	-	1,1
3. Hämophile	95	0	5,4
4. Bluttransfusionsempfänger	27	18	2,6
5. Heterosexuelle Partner von Risikogruppen 1.-4.	40	17	3,2
6. Kinder unter 13 J. (Eltern aus Risikogruppe)	11	5	0,9
7. Nicht bekannt	71	10	4,6
Gesamt	1651	109	

c)

Alter	Zahl der Patienten männlich	Zahl der Patienten weiblich	% von Gesamtzahl
0-1 J.	7	5	0,7
1-9 J.	8	3	0,6
10-15 J.	9	0	0,5
16-19 J.	10	0	0,6
20-29 J.	272	44	18,0
30-39 J.	616	32	36,8
40-49 J.	503	11	29,2
über 50 J.	215	13	13,0
unbekannt	11	1	0,7
Gesamt	1651	109	

d)

Table 1: AIDS cases in the Federal Republic of Germany as reported to National Reference Center for AIDS at the Bundesgesundheitsamt, Berlin; after [AIFO]

Cumulative AIDS cases for the period 1.1.82 - 29.1.88 (a), new AIDS cases by period (b) for the different Länder of the FRG and for West-Berlin.

Cumulative figures by risk groups (c) and by age classes (d).

AIDS cases reported to WHO, by country, as of 31st December 1987

Country/Area - Land/Gebiet	Date of report Datum der Meldung	Number of cases Zahl der Fälle
Africa - Afrika		
Algeria - Algerien	01.06.87	5
Angola - Angola	26.09.86	6
Benin - Benin	18.05.87	3
Botswana - Botswana	10.10.87	13
Burkina Faso - Burkina Faso	30.06.87	26
Burundi - Burundi	15.10.87	569
Cameroon - Kamerun	05.03.87	25
Cape Verde - Kap-Verde	30.04.87	4
Central African Republic - Republik Zentralafrika	31.10.86	254
Chad - Tschad	13.11.86	1
Comoros - Komeren	13.11.86	-
Congo - Konge	13.11.86	250
Côte d'Ivoire - Elfenbeinküste	20.11.87	250
Djibouti - Djibouti	01.10.87	-
Egypt - Ägypten	06.07.87	1
Ethiopia - Äthiopien	04.12.87	19
Gabon - Gabun	06.07.87	13
Gambia - Gambia	16.03.87	14
Ghana - Ghana	25.05.87	145
Guinea - Guinea	12.11.87	4
Guinea Bissau - Guinea Bissau	20.11.87	16
Kenya - Kenia	10.11.87	964
Lesotho - Lesotho	27.11.87	2
Liberia - Leberia	12.06.87	2
Madagascar - Madagaskar	25.04.87	-
Malawi - Malawi	13.11.86	13
Mali - Mali	08.09.87	-
Mauritania - Mauretanien	13.11.86	-
Mauritius - Mauritius	15.09.87	1
Mozambique - Mosambik	08.12.87	4
Nigeria - Nigeria	22.05.87	5
Reunion - Reunion	10.06.87	1
Rwanda - Ruanda	30.11.86	705
Sao Tomé and Principe - Soa Tomé und Principe	01.12.86	-
Senegal - Senegal	13.10.87	27
Seychelles - Seychellen	13.11.86	-
Sierra Leone - Sierra Leone	03.11.87	-
South Africa - Südafrika	10.12.87	93
Sudan - Sudan	23.08.87	12
Swaziland - Swaziland	01.07.87	7
Togo - Togo	13.11.86	-
Tunisia - Tunesien	06.12.87	11
Uganda - Uganda	31.10.87	2 369
United Republic of Tanzania - Vereinigte Republik Tansania	17.10.87	1 608
Zaire - Zaire	30.06.87	335
Zambia - Zambia	09.12.87	536
Zimbabwe - Simbabwe	28.08.87	380
Total - Gesamt		8 693
Americas - Amerika		
Anguilla - Anguilla	31.03.87	2
Antigua and Barbuda - Antigua und Barbuda	30.06.87	3
Argentina - Argentinien	30.09.87	120
Bahamas - Bahamas	16.10.87	163
Barbados - Barbados	30.09.87	52
Belize -Belize	30.09.87	4
Bermuda - Bermuda	30.09.87	75
Bolivia - Bolivien	16.10.87	4
Brazil - Brasilien	27.09.87	2 325
British Virgin Islands - Britische Virgin Inseln	31.03.87	-
Canada - Kanada	14.12.87	1 423
Cayman Island - Cayman Inseln	31.03.87	2
Chile - Chile	30.09.87	56
Colombia - Kolumbien	30.09.87	153
Costa Rica - Costa Rica	30.09.87	39
Cuba - Kuba	16.10.87	6
Dominica - Dominica	30.09.87	5
Dominica Republic - Dominikanische Republik	16.10.87	352
Ecuador - Equador	30.09.87	52
El Salvador - El Salvador	03.10.87	16
French Guiana - Französisch Guinea	16.10.87	93
Guadeloupe - Guadeloupe	30.06.87	51
Grenada - Grenada	16.10.87	7
Guatemala - Guatemala	30.09.87	30
Guyana - Guyana	30.09.87	5
Haiti - Haiti	30.09.87	912
Honduras - Honduras	15.09.87	51
Jamaica - Jamaica	30.09.87	30
Martinique - Martinique	30.06.87	27
Mexico - Mexiko	16.10.87	713
Montserrat - Montserrat	30.09.87	-
Nicaragua - Nicaragua	18.09.87	19
Panama - Panama	30.09.87	22
Paraguay - Paraguay	30.06.87	14
Peru - Peru	30.09.87	44
Saint Christopher and Nevis - Saint Christopher und Nevis	30.09.87	1
Saint Lucia - Santa Lucia	30.09.87	6
Saint Vincent and the Grenadines - Saint Vincent und die Grenadinen	30.09.87	7
Suriname - Surinam	30.09.87	6
Trinidad and Tobago - Trinidad und Tobago	30.11.87	206
Turks and Caicos Islands - Turks und Caicos Inseln	30.06.87	4
United States of America - USA	28.12.87	49 743
Uruguay - Uruguay	30.09.87	14
Venezuela - Venezuela	30.09.87	101
Total - Gesamt		56 958
Asia - Asien		
Bangladesh - Bangladesch	14.04.87	-
Bhutan - Bhutan	14.04.87	-
Brunei Darussalam - Brunei Darussalam	08.09.87	-
Burma - Burma	14.04.87	-
China - China	08.09.87	2
China (Province of Taiwan) - China (Provinz Taiwan)	26.01.86	1
Cyprus - Zypern	01.06.87	3
Democratic People's Republic of Korea - Volksrepublik Korea	09.05.87	-
Eastern Mediterranean Region - Östliche Mittelmeerregion	10.09.87	36
Hong Kong - Hong Kong	17.11.87	6
India - Indien	09.05.87	9
Indonesia - Indonesien	21.04.87	1
Israel - Israel	30.09.87	43
Japan - Japan	14.12.87	59
Jordan - Jordanien	06.12.87	3
Lebanon - Libanon	03.06.87	3
Malaysia - Malaysien	08.09.87	1
Maldives - Malediven	30.06.87	-
Mongolia - Mongolei	30.09.87	-
Nepal - Nepal	09.05.87	-
Philippines - Philippinen	30.10.87	10
Qatar - Katar	09.05.87	9
Republic of Korea - Republik Korea	08.09.87	1
Singapore - Singapur	30.06.87	2
Sri Lanka - Sri Lanka	14.04.87	2
Thailand - Thailand	12.10.87	12
Turkey - Türkei	30.06.87	21
Viet Nam - Viet Nam	08.09.87	-
Total - Gesamt		224
Europe - Europa		
Albania - Albanien	31.08.87	-
Austria - Österreich	30.09.87	120
Belgium - Belgien	30.09.87	280
Bulgaria - Bulgarien	06.10.87	3
Czechoslovakia - Tschechoslowakei	30.09.87	7
Denmark - Dänemark	30.09.87	202
Finland - Finnland	30.09.87	22
France - Frankreich	30.09.87	2 523
German Democratic Republic - DDR	30.09.87	4
Germany, Federal Republic of - BRD	30.11.87	1 588
Greece - Griechenland	30.09.87	78
Hungary - Ungarn	30.09.87	6
Iceland - Island	30.09.87	4
Ireland - Irland	30.09.87	25
Italy - Italien	30.09.87	1 104
Luxembourg - Luxemburg	30.09.87	8
Malta - Malta	30.09.87	7
Netherlands - Niederlande	30.09.87	370
Norway - Norwegen	30.09.87	64
Poland - Polen	30.06.87	3
Portugal - Portugal	30.09.87	81
Romania - Rumänien	30.09.87	2
Spain - Spanien	30.09.87	624
Sweden - Schweden	07.12.87	156
Switzerland - Schweiz	30.09.87	299
USSR - UDSSR	05.08.87	4
United Kingdom - Großbritannien	04.12.87	1 170
Yugoslavia - Jugoslawien	30.09.87	21
Total - Gesamt		8 775
Oceania - Ozeanien		
Australia - Australien	07.12.87	681
Cook Islands - Cook Inseln	08.09.87	-
Fiji -Fidji	08.09.87	-
French Polynesia - Franz. Polynesien	08.09.87	1
Kiribati - Kiribati	26.10.87	-
Mariana Islands - Marianen	05.08.87	-
New Caledonia and Dependencies - Neukaledonien	08.09.87	-
New Zealand - Neuseeland	14.12.87	59
Papua New Guinea - Papua Neu-Guinea	08.09.87	-
Samoa - Samoa	08.09.87	-
Solomon Islands - Salomoninseln	08.09.87	1
Tonga - Tonga	06.10.87	1
Tuvalu - Tuvalu	08.09.87	-
Vanuatu - Vanuatu	08.09.87	-
Total - Gesamt		742
World total - weltweit Gesamt		75 392

Table 2: Global data, after [AIFO]

2.2. Peculiarities of the HIV-infection with Relevance for the Epidemic Dynamics

In this section, I will summarize those features of the AIDS epidemic which set it apart from conventional infectious diseases; these may be considered well known facts by now, as long as they are stated as qualitative features. For those who want to go deeper into the subject I have two recommendations for introductory reading: The book by Koch [Ko] is a good primer; while some of the views of the author are quite controversial the book contains a lot of information and keeps a good balance between popular science writing and presentation of details. In the pages of Nature, a lively debate is going on about many aspects of AIDS, from biomolecular to epidemiological.

The transmission of HIV-infection occurs mainly along well-defined narrow specific channels, namely:

sexual contacts
needle sharing
perinatally from mother onto child;

in addition, there are rare unspecific transmissions like:

visiting prostitutes
unspecific sexual contacts
stitches with infected instruments
import of infection.

Further charateristics are: low infectiosity; a long incubation period (with a mean of about 10 years or even more for the time span from infection to full blown AIDS); a life long infectiosity; AIDS is with certainty lethal, there is no cure at present (and due to the complicated biochemistry and the fact that HIV attacks prominent cell types of the immunoresponse system itself there will probably be none for the foreseeable future). There are many unsymptomatic carriers of infection; the risk of transmission varies considerably with individual habits; the risk of transmission is, in principle, open to influence and change, however, on the level of individual behaviour only. The possibility has to be allowed for that the course of the illness, i.e. incubation time and infectiosity, varies with the different roles individuals with different sexual attitudes play for the transmission. We have to stay open for the influence of different varieties of HIV and of the high variability of the virus.

Not yet clarified is the influence of mode of transmission, viral dosis and cofactors on risk of transmission, incubation period and infectiosity of an infected person as a function of time since infection. If measured virus titers could be taken as a gauge for the latter we would have something like:

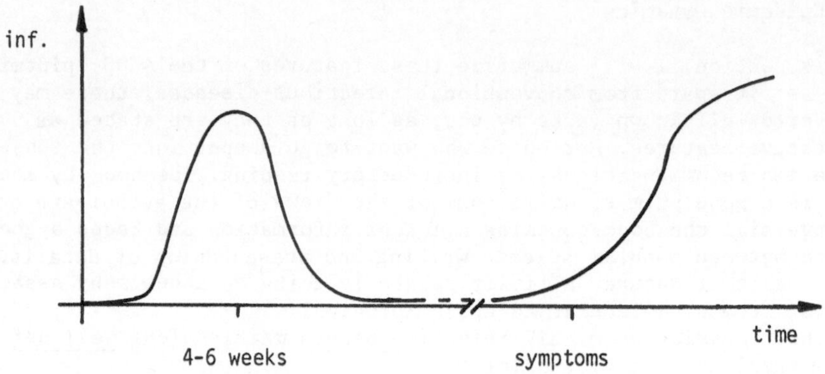

Note that "infectivity" has been given in arbitrary undefined units. There are regional differences, between metropolitan and rural areas say, leading to regional transport of infection.

When we now recall the wide variety in sexual behaviour, not only in activity and practices but especially in the modes of partnership formation (number of partners, duration of partnership, preferences in partner selection), then we have collected the facts an epidemiological model for AIDS has to account for.

3. STANDARD EPIDEMIOLOGICAL MODELS

In this paragraph we will have a short look at standard models in epidemiology, confining the scope to deterministic ones. For a general introduction to the theory of infectious diseases see the booklet by Frauenthal [Fr] and the thorough textbook by Bailey [Ba]; the monograph by Hethcote and Yorke [HY] is devoted to sexually communicable diseases.

3.1. "Simple Epidemics"

The simplest situation which might be considered a model for the transmission of a infectious disease consists in a population of fixed size N and a number $I(t)$ of infected individuals; they are infective at the same time. The rate of transmission is supposed to be proportional to the number of infected and the number of the yet uninfected, hence

$$\frac{dI}{dt} = \lambda\ I(N - I)$$

and, with simple scaling, we arrive at the logistic equation

$$\frac{dx}{dt} = \lambda\ x(1 - x) \quad .$$

This situation has been termed "simple epidemics" [Ba] .

3.2. 2-Component Simple Epidemics

If we consider 2 species, denoted by M and F, with $\{M,F\}$ = {male, female} for heterosexual transmission or = {man, fly} in the case of malaria, where transmission occurs only between species and not inside one, we arrive at

$$\frac{dI^M}{dt} = p_1 I^F (N^M - I^M)$$

$$\frac{dI^F}{dt} = p_2 I^M (N^F - I^F) .$$

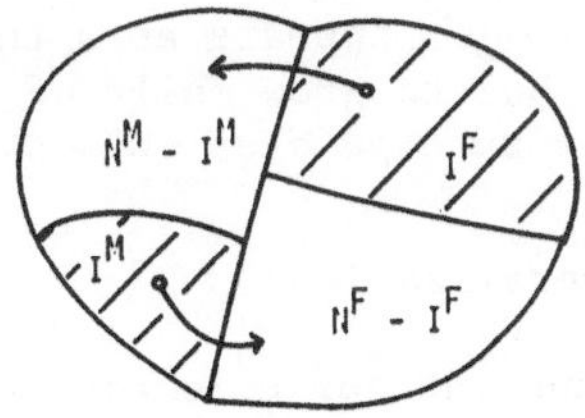

Again, the population is constant in size, $N = N^M + N^F$, with respective numbers of infected $I^M(t)$, $I^F(t)$ in the two species; appropriate scaling leads to the Lotka-Volterra equation

$$\frac{dx}{dt} = p_1' \, y(1 - x)$$

$$\frac{dy}{dt} = p_2' \, x(1 - y) .$$

3.3. "General Epidemics": IRS-models

We now come back to one species, but we add to the mechanism of infection the possibility of removals of infectives; thus we have 3 states: individuals who are initially susceptible may become infected and are then removed, be it by death or through recovery together with immunity. Thus we have:

$$N = I(t) + R(t) + S(t) = \text{const.}$$

$$\frac{dS}{dt} = - r \, S \, I \qquad , \quad r = \text{infection rate}$$

$$\frac{dI}{dt} = r \, S \, I - \gamma \, I \qquad , \quad \gamma = \text{removal rate}$$

$$\frac{dR}{dt} = \gamma \, I \quad .$$

An interesting fact is recognized when one looks at the flow of the number of infected with the number of susceptibles:

$$\frac{dI}{dS} = \frac{r\,S\,I - \gamma\,I}{-\,r\,S\,I} = -\,1 + \frac{\rho}{S} \quad , \quad \rho = \frac{\gamma}{r}$$

$$I(S) = N - S + \rho \ln \frac{S}{S_o} \quad (R_o = 0) \quad .$$

Here, $\rho = \frac{\gamma}{r}$ appears as a threshold value since, for an epidemic to occur, I has to grow while S is falling hence we need an initial number of susceptibles $S_o > \rho$.

3.4. Generalizations

We may augment our epidemic dynamics by dividing the population into different groups distinguished by different values of the dynamical parameters:

$$N = \sum_{i \in I} N_i \ ; \quad I_i \, , R_i \, , S_i \, , r_{ij} \, , \gamma_i \quad .$$

In the case of AIDS one would consider in a first approach (arrows indicate major routes of transmission):

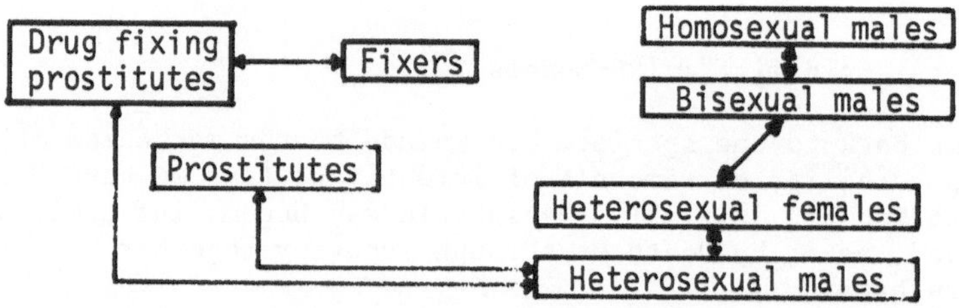

One rapidly ends up with large systems of differential equations. A typical model [WSK] of this type with essentially the above group structure and taking into account the different stages of the HIV-infection contains about 50 variables. A further subdivision into age-classes leads to a system of approximately 2000 coupled nonlinear differential equations.

A first step into another direction one may go into in generalizing the classical models of epidemiology is taken by Dietz [Di1,Di2] and Dietz and Hadeler [DH]: They consider a heterosexual population and treat the formation of partnership by introducing extra groups representing pairs with both partners uninfected, with the male infected and the female uninfected and so on. Thus the group structure of this model looks as follows (with arrows showing possible migration between groups):

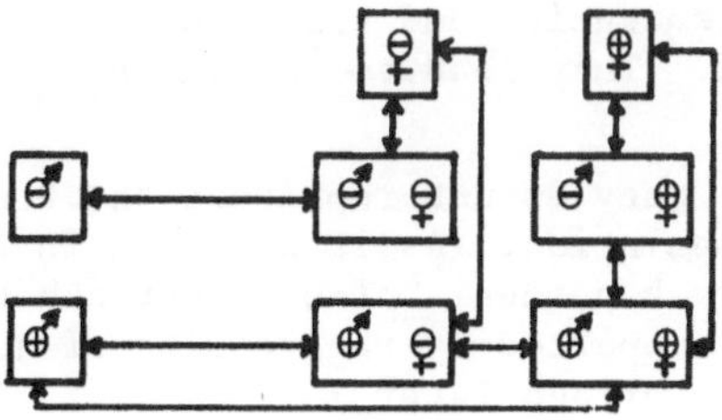

This is clearly not yet a model for AIDS; however it goes into the right direction since it tries to incorporate into the framework of the standard differential equation models a description of the dynamics of partner selection which is a most important aspect for a model for AIDS. The intention is to augment it, in a next step, by the introduction of age classes and age correlation in the formation of partnerships.

Of course, there exist other approaches in epidemiology; as two important ones let us mention stochastic versions of the differential equation models and discrete time stochastic models the prototypes of which are coming under the name of chain-binomial models. For a few other attempts at modelling the spread of AIDS, I would like to refer to [An], [Kn] and [HS].

3.5. Critique of Standard Models

As a preparation for the motivation of our own approach in the next paragraph, I will collect what we regard as the principal weaknesses of the standard epidemiological models when applied to HIV-infection.

- They describe, through their mass-action type dynamics, permanent potential contact between all members of the groups involved. (This is certainly fine for malaria, but it is inappropriate in the case of HIV. It can be overcome by introducing artificial groups, one for each possible type of pairing and for each kind of other partnership, if nonmonogamous behaviour is to be included.)
- The behaviour is uniform over each group. (They can take account

of nonuniform behaviour only at the prize of division into subgroups, i.e. higher dimensionality of their systems of differential equations. The numerical effort grows multiplicatively.)

- They cannot directly take into account varying numbers of partners. (Again this is possible only by introducing extra groups.)
- Time delays, age dependencies and time dependent rates are not easily incorporated. (They lead to partial differential equations in general.)
- Generally speaking, they do not represent the true contact structure of the population. (As it was said in the introduction, it is the main hypothesis of our appraoch that this matters. This will be supported by the results of actual computer simulations to be described later.)

4. HEURISTIC DESCRIPTION OF OUR APPROACH

In [BBK] we proposed to use random graphs and discrete time stochastic processes to model the epidemic dynamics of sexually transmitted diseases. Here I will describe this approach in heuristic terms, trying to clarify the conceptual layout. This paragraph should be viewed in conjunction with the next one, which will describe it in mathematical terms, and with section 7, which will take a down to earth point of view of practical computer simulations. It is important to combine computer simulations of possibly quite intricate model situations, coming as close as possible to reality, with analytical treatment of simplified situations, helping to understand the principal aspects of the mathematical structures exploited. While first results from computer simulations will be presented in this contribution and in [Bo], see [BBK],[B1],[BK1],[BK2] for first analytical results.

4.1. The Basic Idea

The basic idea of our approach is to directly simulate the transmission of infection between individuals of a model population which represents the real contact relations in structure and dynamics. Evidently, not every detail about the true contact structure inside a society or all the determinants of individual sexual acts can be known nor are they needed from an epidemiological point of view. Hence the natural attitude is to use probabilistic concepts for both the contact structure and the dynamics of transmission unrolling on it. Taking the point of view that real life itself is not a deterministic process and that, furthermore, a large portion of our personal preferences is of no relevance to the epidemic dynamics, we might use the slogan: probability enters into our model as it enters into real life. Remains the task of identifying those parameters which represent that portion of people's preferences and attitudes that is of epidemiological relevance.

4.2. The Fixed Structure

The frame of our modelling approach will be presented by listing the conceptual entities, their dynamical function and the mathematical construct used to represent them. By the way, the set of simulation programs is structured along the same lines.

Configuration:

This is the *contact structure* consisting in the individuals, divided into groups according to their sexual behaviour, together with the mesh of their contacts with their sexual partners. It is represented by a graph, whose vertices correspond to the individuals and the edges to their sexual partnerships during a given time interval; since this graph is only specified by a set of global parameters, we are actually considering random graphs.

Dynamics:

The process of *transmission* evolves in discrete time steps (of 1 day) and consists in the eventual realization of one of the potential sexual contacts for each individual plus the eventual transmission of infection along this realized contact. These events occur according to prescribed probabilities and the spread of infection is described by a discrete time stochastic process evolving on the underlying graph whose state space characterizes the state of infection of the population.

Reconfiguration:

The *change of contact structure* takes place on an intermediate time scale (of a few months, say) when individuals may eventually acquire new partners. This corresponds to the dynamics continuing to evolve on a different realization of the random graph or, in case we let the parameters specifying the graph change too, on a realization of a different random graph; the new graph inherits the infection state from the old one, of course.

Measurements:

In order to *determine the state* of infection at the end of the time for which the dynamics was running (of the order of years) or at any time in between, the whole state of the population will be saved and processed by independent measurement programs which can produce a variety of epidemiologically interesting figures.

Protocols:

The detailed structure of our model is amenable to a more *refined analysis* if we produce, during the run of the dynamics, protocols registering charateristic events. This will allow to understand the influence of the graph structure on the course of the dynamics or to follow the fate of single individuals which is especially interesting for the simulation of prevention strategies.

4.3. Variable Elements

What is fixed in our model is merely the conceptual frame while all the specific aspects remain variable and highly adaptable. To mention a few:

- All epidemiological parameters enter via freely definable distribution functions.
- The number of individuals, of groupes, of stages of infection and illness and the maximal number of partners are variable. (Number of partners refers here to the time between two reconfiguration steps and is limited only by practical programming considerations.)
- Qualitative features of the dynamics and of the algorithms to produce the graphs used in the simulations may be changed. (Actually, there are subtle points involved here, related to the proper definition and realization of random graphs.)
- Nonsexual transmission modes can be implemented as "external channels" by setting additional infections according to certain probability distributions.
- We can allow for changes in behaviour: at any time with respect to the dynamics, at reconfiguration steps with respect to the configuration.

5. RANDOM GRAPH EPIDEMICS

A space of random graphs, or a random graph for short, is the set of all graphs fulfilling a collection of conditions which then specify this random graph. For an introduction to the theory of random graphs see [Bs].

5.1. The Underlying Graph Structure

We will consider labeled graphs with vertex set

$$V = \{x_1, \dots, x_N\} \quad ,$$

divided into groups

$$V = \bigcup_{i \in I} V_i \quad , \quad |V_i| = N_i \quad , \quad \Sigma N_i = N \quad .$$

For any set A , $|A|$ and $\#\{a \in A\}$ denote its cardinality. As possible specifications for random graphs we might consider:

$$n : \mathbb{N} \to \mathbb{N}$$
$$n(z) = \#\{x_i \in V \mid d(x_i) = z\}$$

$$n_i : \mathbb{N} \to \mathbb{N}$$
$$n_i(z) = \#\{x_j \in V_i \mid d(x_j) = z\}$$

$$n_{ij} : \mathbb{N} \to \mathbb{N} \times \mathbb{N}$$

$$n_{ij}(z) = \left(\# \{x_k \in V_i \mid d_j(x_k) = z\}, \# \{x_l \in V_j \mid d_i(x_l) = z\} \right)$$

$$f_{ij} : \mathbb{N} \to \mathbb{R}_+$$

$$f_{ij}(z) = \text{prob}\{ d_j(x_k) = z \quad \text{for} \quad x_k \in V_i \}$$

$$d_{ij} \in GL(|I|)$$

$$d_j(x_k) = d_{ij} \quad \forall x_k \in V_i$$

where

$$d(x_i) = \sum_{y \in V} \Phi(x_i, y)$$

$$d_j(x_k) = \sum_{y \in V_j} \Phi(x_k, y)$$

and Φ is the adjacency matrix of the graph. The corresponding spaces of all graphs having the above degree distributions will be denoted by G^n, G^{n_i}, $G^{n_{ij}}$, $G^{f_{ij}}$, $G^{d_{ij}}$, respectively, and will be endowed with a probability space structure by equally weighting all graphs:

$$\text{prob}(G) = \frac{1}{|G^{\cdots}|} \quad \forall\ G \in G^{\cdots} \quad .$$

To illustrate possible further refinements in the specifications let us look at three examples:

i) Consider $x_i \in V$ and connected subsets $Cl(x_i,k,n) \subset V(G)$, "k-n-clusters", with:

$$|Cl(x_i,k,n)| = n$$

$$\sum_{\substack{x \in Cl(x_i,k,n) \\ y \in V(G) \setminus Cl(x_i,k,n)}} \Phi(x,y) = k$$

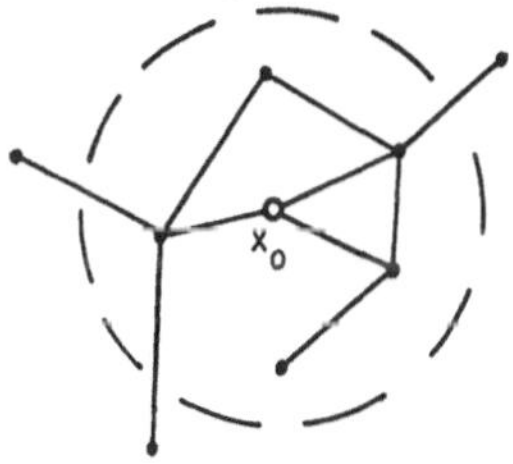

$Cl(x_0,3,6)$

Specify the size-distribution and outer-edge-distribution of clusters centering around individuals of a given group.

ii) Consider $x_i \in V$, look at cycles, i.e. closed chaines of edges. Specify the distribution of cycle length.

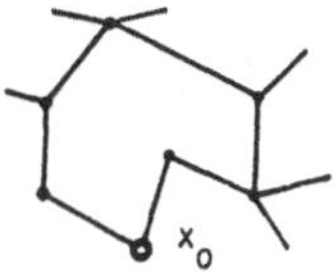

cycle of length 7

iii) Consider $x_1, x_2, x_3 \in V_1$, $y_1, y_2, y_3 \in V_2$. Let $d(x_1,y_3) = k$ where $d(\cdot,\cdot)$ denotes the distance between two vertices, i.e. the minimal number of edges connecting them. Suppose $d(x_1,x_2) = \ell$; what can we know about

$$\text{prob}\,\{d(x_2,y_3) = k'\}$$

for $k' < k + \ell$?

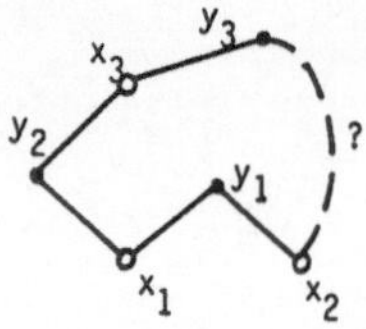

The specifications we select to define our random graph spaces have to reflect available or accessible sociological data about sexual behaviour. Since these are not the specifications mostly considered in random graph theory there remain a lot of open questions about the properties of these random graph spaces. Important problems are: the size of the spaces; properties of typical graphs and the related question of whether an actually constructed graph is a good representative; the relation between different random graph spaces and especially with known models in random graph theory. Let us remark that from a practical point of view we can be quite satisfied with approximate realizations of the specified graphs.

5.2. Epidemic Dynamics as Discrete Time Stochastic Processes

We introduce a state vector for each vertex in the graph

$$\vec{\chi} : V \to \{0,1\}^K =: S$$

with the interpretation that

$$\chi_k(x_i) = 1 \sim x_i \begin{cases} \text{is infected} \\ \text{is seropositive} \\ \text{knows about infection} \\ \text{has LAS} \\ \text{has ARC} \\ \text{has AIDS} \\ \text{is dead} \end{cases}, \quad k = \begin{cases} 1 \\ \vdots \\ K \end{cases}.$$

The transitions $\chi_k \to \chi_{k+1}$ are governed by distribution functions (for seroconversion latency time, incubation period ...) and occur autonomously at each vertex.

The dynamics proper, i.e. the spread of infection, is described as a discrete time stochastic process over the relevant space of random graphs with state space S^N.

The time step τ_0 is 1 day, and S^N is the infection state of the whole population. The occurence of events, i.e. the act of transmission of a single infection, is governed by the following probabilities:

$\forall\, x_i \in V :$ $\quad p_a(x_i)$ ~ overall activity of an individual

$\forall\, x_i \in V,\ x_j \in K(1,x_i) :$ $\quad p_s(x_i,x_k)$ ~ relative probability to select one of the partners

$\forall\, e_i \in E,\ x_j,x_k \in e_i :$ $\quad p_i(x_j,x_k)$ ~ infection probability per single contact

Here, $K(1,x_i) = \{y \in V \mid d(y,x_i) = 1\}$ is the 1-neighbourhood of x_i, i.e. the set of its partners, and E denotes the set of edges of G. The p_α are given via group specific distribution functions $f_{\alpha,i}, i \in I$. These may as well depend on the state of the individuals

$$f'_{\alpha,i} = g(\vec{\chi}(x_j))\cdot f_{\alpha,i} \qquad \text{for } x_j \in V_i \quad ;$$

strictly speaking, they always do insofar as the occurence of events does depend on the boolean variable $\vec{\chi}$, but numerical values may depend on $\vec{\chi}$ too. Further dependencies can be incorporated as well, e.g. on the state of the neighbourhood of an individual or on the time elapsed since the transition into a certain state.

Remark that, equivalently, we can look upon the dynamics as the random graph process

$$G_{t_n}^{(I)} \in \left(G^{n_{ij}}\right)^{\mathbb{N}}$$

generating the infection subgraph.

5.3. Some particular cases

5.3.1. The complete bipartite graph K_{N^M,N^F} – (heterosexual transmission in a completely permissive society): Consider a bipartite graph $V = V_M \cup V_F$ with N^M males and N^F females, $N^M \leq N^F$, and I_o^M, I_o^F given. Count time by n ; in each time step edges are chosen anew, such that a complete pairing of N^M into (a subset of) N^F is achieved. Then we have for the number of infected males and females

$$\mathbb{E}\left[I_{n+1}^{M,F}\right] = \mathbb{E}\left[I_n^{M,F}\right] + p_{1,2}\ \mathbb{E}\left[{}^{*}I_n^{F,M}\right]$$

where $p_1(p_2)$ is the transmission probability per contact from female to male (from male to female) and the expected number of contacts between infected females (or males) and noninfected males (or females) is given by

$$\mathbb{E}\left[{}^{*}I_n^{F,M}\right] = N^M\ \frac{\mathbb{E}\,[I_n^{F,M}]}{N^{F,M}} \cdot \frac{N^{M,F} - \mathbb{E}\,[I_n^{M,F}]}{N^{M,F}} \qquad .$$

Using

$$x_n = \frac{\mathbb{E}[I_n^M]}{N^M}\ ,\quad y_n = \frac{\mathbb{E}[I_n^F]}{N^F}\ ,\quad q_1 = p_1\ ,\quad q_2 = \frac{N^M}{N^F}\, p_2$$

we obtain the time discretized Lotka-Volterra equation

$$\begin{aligned} x_{n+1} &= x_n + q_1 y_n (1 - x_n) \\ y_{n+1} &= y_n + q_2 x_n (1 - y_n) \quad . \end{aligned}$$

In this sense we might say that standard ordinary differential equation models are special, namely complete n-partite graph, cases of our model.

5.3.2. Regular graphs, $d(x_i) = r \ \ \forall\, x_i \in V$: A lot is known about graphs in $G^{\text{r-regular}}$; for more details about the following, see [BBK,B1] .

r = 2 : The whole graph necessarily splits into cycles; the distribution of the order of the cycles can be calculated.

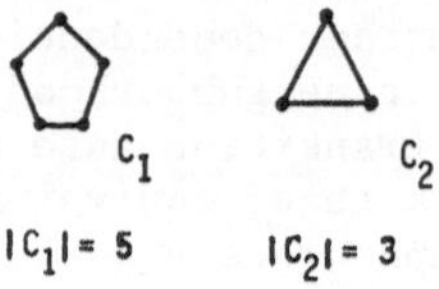

r = 3 : With special assumptions, the infection subgraph acquires a tree-like structure and, introducing two auxiliary variables, the dynamics can be cast into a 3-dimensional iterative map:

$$\begin{aligned} I^{n+1} &= p(I^n + U^n) - \frac{(I^n + U^n)(I^n + U^n - 1)}{X^n + U^n - 1} \\ U^{n+1} &= I^n \\ X^{n+1} &= X^n - I^n \quad . \end{aligned}$$

Thus we can study special models in random graph epidemics by studying related iterative maps.

6. LINKS TO RANDOM GRAPH THEORY AND TO PERCOLATION THEORY

As I indicated in the introduction, this paragraph is rather daring; I would like it to be seen as just a collection of vague remarks intended to arouse your interest.

In random graph theory the following 3 models, i.e. spaces of random graphs, are considered most:

$$G^{N,p}\ ,\quad G^{N,M}\ ,\quad G^{N,\text{r-regular}}$$

where

$N = \#\text{vertices} = |V|$

$M = \#\text{edges} = |E|$

p = probability of an edge being realised, i.e. for $G \in G^{N,p}$: $G \subset K^N$ and K^N is the complete graph of order N

r-regular : $d(x_i) = r \quad \forall\, x_i \in V$.

The models $G^{N,p}$ and $G^{N,M}$ are almost interchangeable provided $M = p\binom{N}{2}$ since $\binom{N}{2} = \#$ pairs of vertices.

A random graph process is a Markov chain $\tilde{G} = (G_t)_{t=0}^{\infty}$ with state space G , a random graph space. One should think of $\tilde{G}$ as a graph growing by acquiring new edges. One studies threshold functions and critical times for properties to appear, or say: hitting times for properties. It might be worthwhile to take this point of view in looking at the infection subgraph.

As a result with relevance for our simulations let us cite a theorem (conf. [Bs], ch. X.3.):

> Let $r \geq 3$, $\varepsilon > 0$ be fixed and let $d = d(n)$ be the least integer satisfying
> $$(r-1)^{d-1} \geq (2+\varepsilon)\, r\, n \log n \quad .$$
> Then almost every r-regular graph $G \in G^{n,r\text{-regular}}$ of order n has diameter at most d .

Here the diameter is defined as the maximal distance between any two vertices, measured in the natural metric on the graph, i.e. by counting the number of edges along the shortest connecting path. Numerically, we have for d :

n \ r	3	4	5	6	7
10000	20	13	10	9	8
20000	21	13	11	10	9
50000	22	14	12	10	9
100000	23	15	12	11	10
200000	24	16	13	11	10
500000	26	17	13	12	11
1000000	27	17	14	12	11

Actually generated graphs tend to have diameter lower than this bound d. Looking at d as the relevant length scale for the spread of infection, we are assured that scaling problems due to the limited size of the model population will not be too serious.

In percolation theory, bond percolation could be viewed as the model $G^{N,p}$, where V = periodic lattice. Most of percolation theory is lattice percolation; however, in general, our graphs cannot be (sensibely) embedded into $\mathbb{Z}^d$. Nevertheless, the fact that percolation thresholds depend strongly on the lattice structure supports the view that the spread of epidemics should strongly depend on the contact structure, while "statistical mixtures" of "pure" graph structures might wash out sharp transitions.

Epidemic models might be regarded as site percolation problems on (a representative sample of) the relevant graphs where edges are loaded with the corresponding transmission probabilities.

An important tool in percolation theory is renormalization; we could hope to learn something about the scaling properties of our models using renormalization techniques. However it is quite unclear how a renormalization transformation could be defined on an irregular graph.

7. SIMULATING THE EPIDEMIC DYNAMICS OF THE HIV-INFECTION

The set of computer programs used for the simulation is structured corresponding to the notions introduced in section 4.2. We have independent programs to generate a graph, to run the dynamics, to perform reconfigurations and to do the measurements. This gives flexibility in using different versions of the programs for the distinct tasks. Thus we can compare different algorithms for constructing a graph while the "same" dynamics is evolving on the different graphs produced. Or we can produce different realizations of the dynamics over the same graph. Different kinds of specifications for the random graph are allowed for through different graph generating programs, while the social parameters determining the graph structure on the one hand and the social and medical parameters determining the dynamics on the other are input variables.

While we like to stress the flexibility of our approach and its adaptability to a wide range of epidemiological assumptions, we have identified two sets of rather simplistic model assumptions to start the simulations with. We dubbed them the "simple model" and the "extended model". The intention is to use the first to explore the spectrum of results our modelling approach will produce under variation of details; the latter should already give a serious approximation to reality.

7.1. The Simple Model

The simple model comprises seven groups which are interpreted as:

1: homosexual males
2: bisexual males
3: heterosexual males
4: heterosexual females
5: heterosexual females having also contact to bisexual males
6: male intravenous drug users
7: female intravenous drug users

The population size is N = 20000 , no reconfiguration will be performed, the dynamics runs over about 20 years. Note the unrealistic feature that we have fixed partnership and no natural vitality dynamics for this time span. The number of partners, the contact frequencies and the risk of infection per contact are group dependent, but homogeneous over each group. The values of the parameters for a reference set are listed in table 3. The results presented in the next section refer to this set of assumptions.

NINDI = 20000 , NGROUP = 7								
group		1	2	3	4	5	6	7
no. of individuals		664	136	9526	9504	136	17	17
max. no. of partners		5	5	4	2	2	4	5
d_{ij} =	1	4	1					
possible number	2	3	1			1		
of partners from group j	3				2	1		1
for individual in group i	4			2				
	5		1	1				
	6						2	2
	7			1			2	2
p_i^a = total activity		.490	.490	.200	.200	.200	1.00	1.00
p_{ij}^s =	1	.200	.200					
relative contact frequency	2	.190	.190			.240		
of individual in group i	3				.500	.500		.050
with partner from group j	4			.500				
	5		.500	.500				
	6						.250	.250
	7			.500			.125	.125
p_{ij}^t =	1	.020	.020					
transmission probability	2	.020	.020			.003		
per contact between	3				.003	.003		.003
infected individual in	4			.001				
group i and uninfected	5		.001	.001				
partner from group j	6						.100	.100
	7			.001			.100	.100

state	infec	spos	LAS	ARC	AIDS
duration	67	1825	600	300	140

Table 3: reference set of parameters

7.2. The Extended Model

The group structure of the extended model is as above, augmented by:

8: prostitutes
9: prostitutes who are also intravenous drug users
10: hemophiliacs .

The most important difference from the computational point of view will be that we use different scaling for the different groups. Note that we have small highly active groups and large less active ones. Using different scaling factors from the real population onto the model population, will allow to more accurately model the small active groups and will thus improve statistics and memory economy. Of course, we have to carefully adjust the number of edges and of realized contacts at the interfaces between groups. Second, using a purely technically motivated further splitting of the groups into subgroups to accomodate the strongly varying number of partners will allow to make more efficient use of memory. With a certain scaling scenario we will be able to typically use 2000000 virtual model individuals with 100000 real model individuals (and 16 MByte of memory).

The extended model allows for unspecific choice of partners, i.e. above the graph structure; so, we will treat as external channels:

visits of prostitutes
very highly promisquous behaviour ($\gtrsim$ 20 partners)
blood transfusion, needle stitch injuries
import of infection .

It will also include:

age structure, vitality dynamics, age dependence of partnerchoice
all parameters as group specific distributions
reconfiguration of the graph
change in behaviour .

7.3. Observables

Measurements are performed by analyzing the state of the population at a given time. As observables we can use the standard observables of epidemiology like overall number of infected, number of infected in one group and any derived quantity like doubling time. More interesting are observables which use more of the detailed information residing in the graph. We can study "conditional observables" like the number of infected amongst those whose number of partners exceeds a certain value or the number of infected in a low risk group as a function of their distance to a high risk group. We can clarify properties of infection paths like their course and typical length. (This cannot be done exhaustively, since they have typically a tree-like structure; however we can sample them or use backtracing). Finally there is the large set of questions related to the characterization of the graph structure; e.g. determining the quality of the approximate realizations of the prescribed graphs or analyzing the cluster properties of both the contact graph and the infection subgraph.

8. FIRST RESULTS FROM THE SIMPLE MODEL

A selection of first results from the computer simulations with the "simple model" is presented here. Note that the parameters are not supposed to be a best possible choice; rather, at this stage of our work, they have been chosen such that they are not completely off the truth while keeping the computer resources needed for a single run modest in order to be able to explore by various runs the potentialities of our model. Furthermore they should be as simple as possible in order to have clear comparison between results produced from varied parameter values. Thus we have a small population size, rather short duration of the infection states and constant parameter values.

The gross features of the structure of our model population should by now become clear from a look at table 3. However, there is one point to be explained. The maximal number of partners for group 3 (heterosexual males) is 4; these might be up to 2 heterosexual female partners (group 4), 1 female partner with contact to a bisexual male (group 5) and 1 drug fixing prostitute (group 7). But due to the small size of groups 5 and 7, most of the heterosexual males have just 2 female partners from group 4. The numbers of individuals in the groups have been chosen such as to reflect the proportion of these groups in society, while group 5 has been separated from group 4 for convenience and its size set to coincide with the size of group 2. (Note that the relative contact frequencies $p_s(x_i, x_k)$ will be determined from the input parameters p^s_{ij} by first setting them to $p_s(x_i, x_j) = p^s_{ij}$ and then renormalizing them to reflect the true number of partners such that

$$\sum_{x_j \in K(1,x_i)} p_s(x_i, x_j) = 1.)$$ Here, an unrealistic feature emerges since groups 6 and 7 are too small in size to have a meaningful dynamics evolve on them. They will rapidly become all infected and, apart from very few "visits of prostitutes" by members from group 3, are disengaged from the dynamics. They will not be included in the presentation of results.

What can we expect? Given that the epidemics starts amongst the homosexuals, inside groups 1 and 2, due to high activity, high transmission probability and the dense mesh of contacts, infection will spread fast. Transport from group 2 via group 5 into the heterosexual population will follow the slower heterosexual transmission probability. Inside the purely heterosexual population, groups 3 and 4, the early infected will be males infected via contacts with group 5, while females will take over due to the higher transmission probability from male to female when there is an autonomous dynamics inside the heterosexual population. The graph will look essentially as follows: As a core, there is a dense mesh of contacts amongst homosexuals interwoven with contacts to bisexuals; with

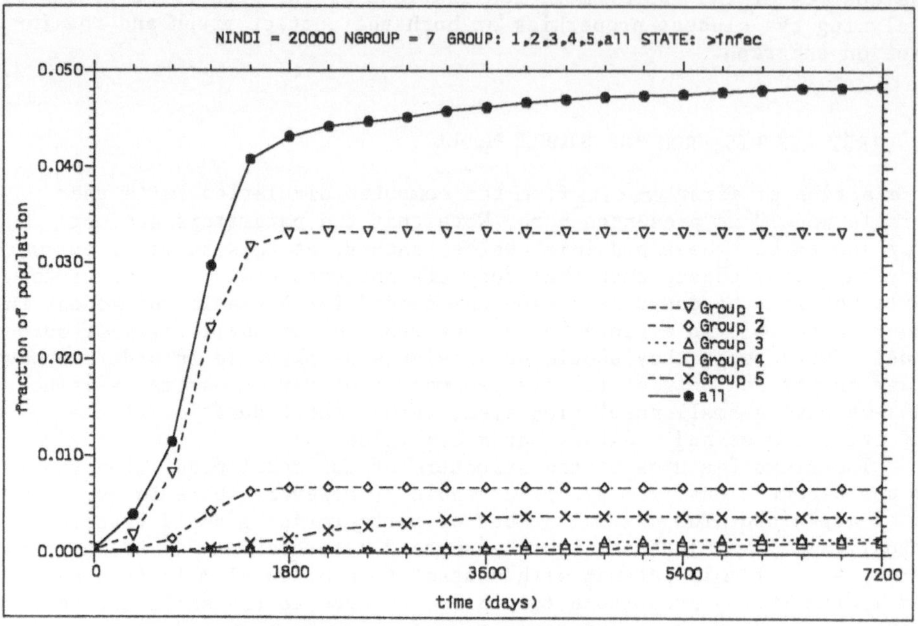

Fig. 1: Reference set of parameters

the bisexuals and their female partners as connectors there are one-dimensional chains of heterosexual contacts emerging from the core due to the fact that heterosexual males and females have just 2 partners from the opposite sex; in addition there are other heterosexual chains or cycles not connected to the core.

The results are presented as plots of the number of infected over time; since the time spent in the different stages of infection is fixed the figures for other states follow with a constant time shift. The number of infected for the different groups are given as fraction of the group size or as fraction of the whole population. For all runs, as initial condition, there have been 6 infected individuals in group 1 and 1 in group 6. Figure 1 shows the result for the reference set of parameters over a time of approximately 20 years. The dynamics inside the homosexuals and the bisexuals is essentially the same as it is seen in figure 2. Increasing both the male to female and the female to male heterosexual transmission probability by a factor of 10 (figure 3) reveals the linear growth on the heterosexual chains. Figures 4 to 6 show

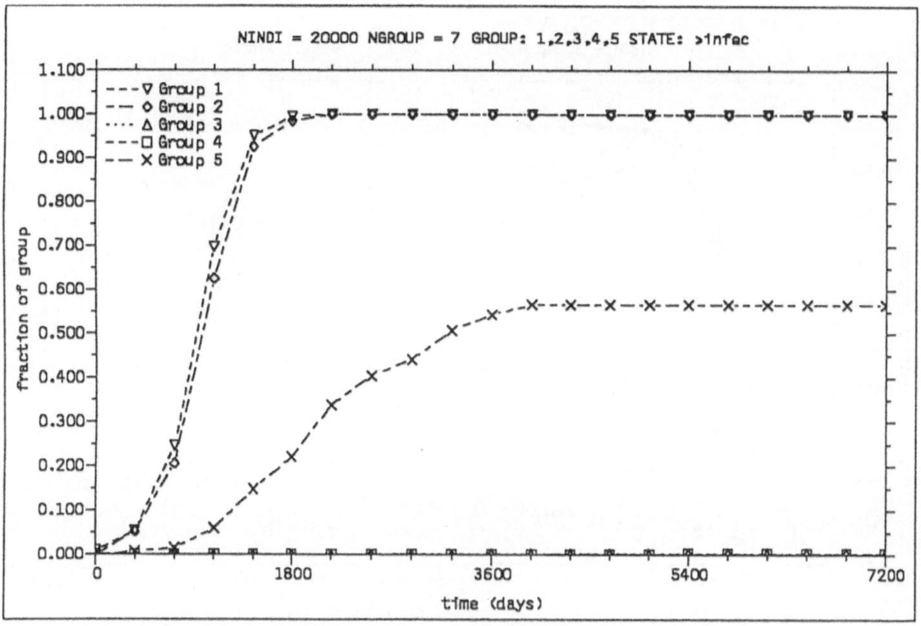

Fig. 2: Reference set of parameters

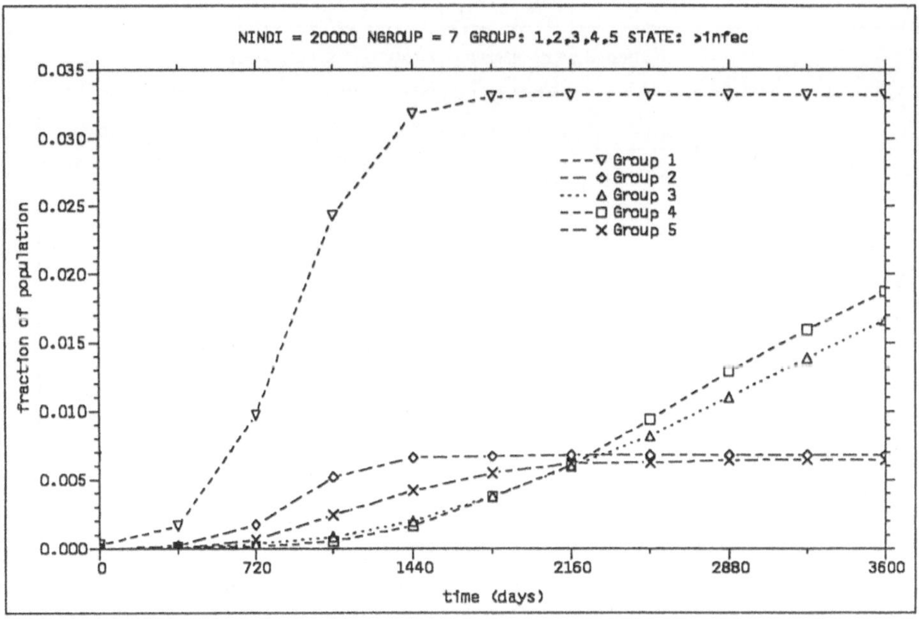

Fig. 3: Heterosexual transmission probability increased by factor 10

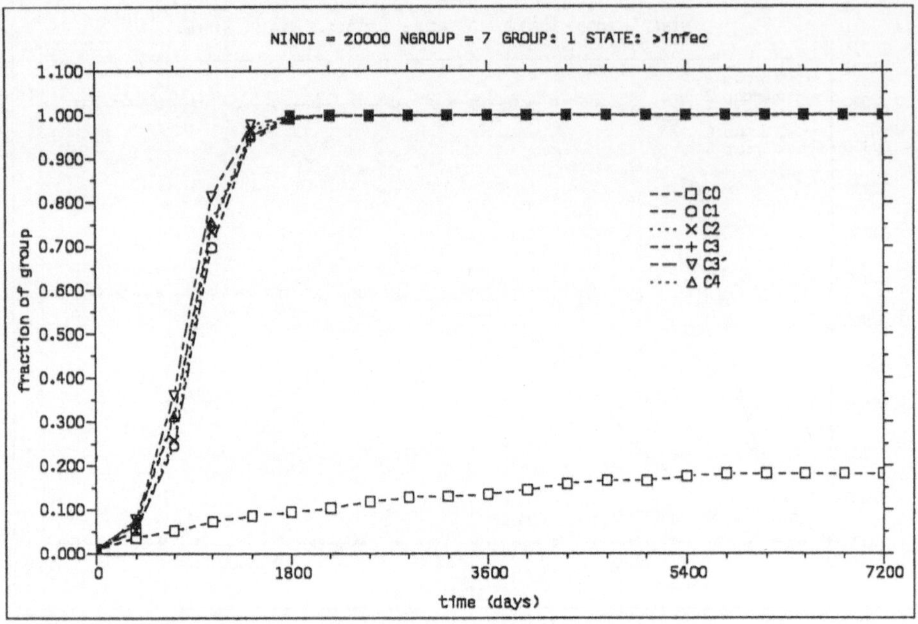

Fig. 4: Comparison of graph algorithms, group 1

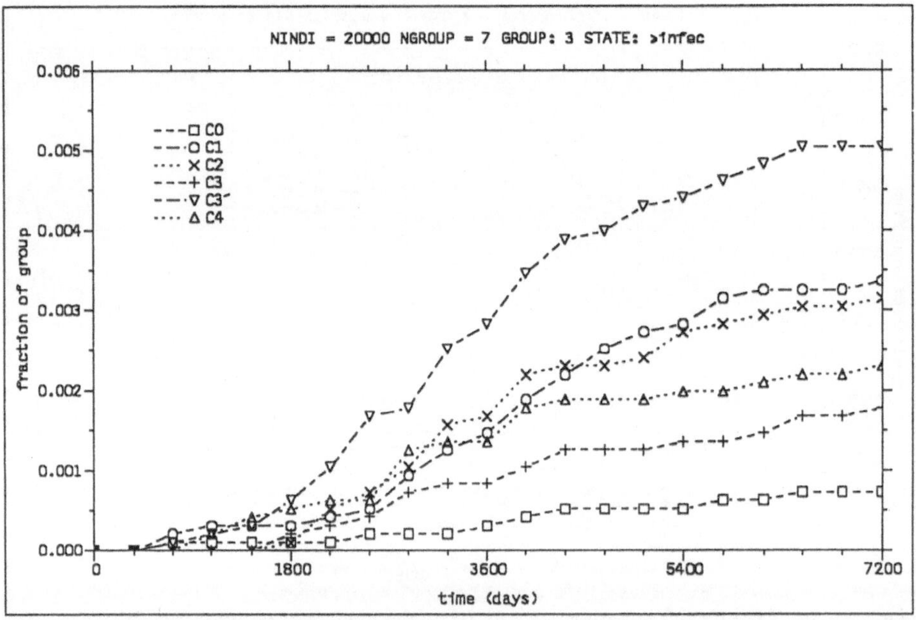

Fig. 5: Comparison of graph algorithms, group 3

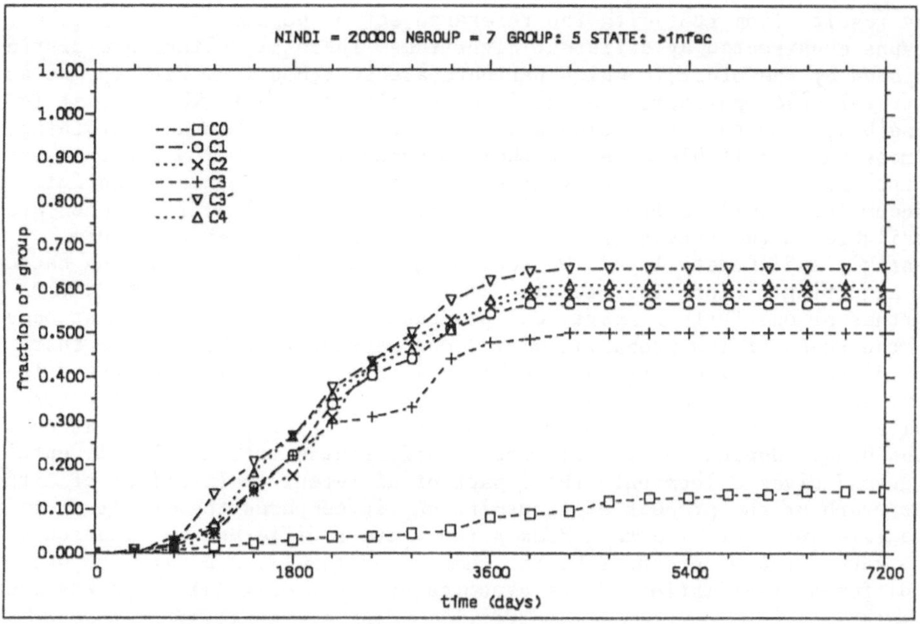

Fig. 6: Comparison of graph algorithms, group 5

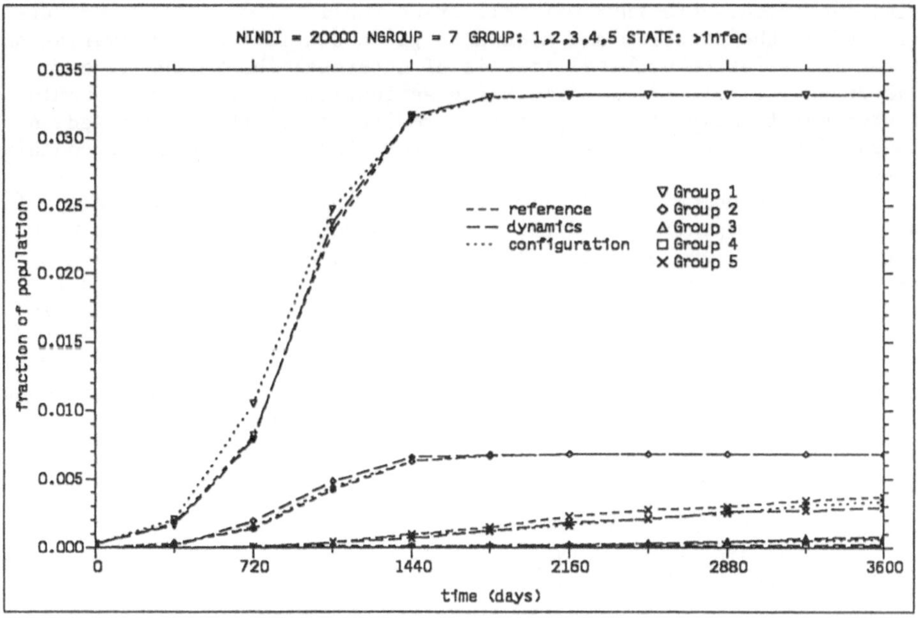

Fig. 7: Reference set of parameters, different realization of configuration and of dynamics

the results from runs with the reference set of parameters on different graphs constructed by different algorithms. These algorithms are distinguished by the order in which partners are searched for. With C∅, as an extremal case, partners are acquired in a linear order, whenever it is possible, such that the adjacency matrix consists of blocks containing almost only small blocks along their diagonal. The other algorithms incorporate random search; only after a certain number of unsuccessful random trials will a linear search be performed until no free bonds are available in the target group. They differ in the number of random searches and in details of the search function leading to varying chances of acquiring a close (w.r.t. the linear order given by the labelling) partner or one further apart. C3' produces the graph which is most open in the sense of low probability of close partner search. Observe that the dynamics on the large active group 1 is not sensitive against different graph structures (except in the extreme case of a block structure), while the efficiency of transmission into the heterosexual population does highly depend on it; for group 5 different saturation levels entail. Figure 7 gives a look onto the impact of different realizations of either the graph or the process of transmission; it compares the result from the reference set with that from a run where a different realization of the random graph, produced by the same algorithm C1, and with one where a different realization of the dynamics over the original graph was used.

9. FINAL REMARKS

Further simulations in the context of the simple model have been and are being performed. More runs with different realization of the configuration and of the dynamics are needed to get a statistics; more variation of the transmission probabilities is of interest; the influence of a time dependent infectiosity including an initial latent period should be studied; finally, the graph structure itself has to be analysed. A general discussion of the results from the simple model will be contained in [Bo].

A major problem in modelling the spread of HIV-infection is the fact that crucial parameters, medical as well as social, are not well known or not in sufficient detail. Available reports about sexual behaviour are so old that one has to suspect that they are outdated or tend to present fancy details of no epidemiological relevance at all; usually data are presented in a stratified format where a matrix of correlated values is needed (for example, data about number of partners and sexual activity gathered over the whole population each are of little use when the number of partners of the highly sexually active is called for by a model detailed enough to dynamically distinguish between such cases). Key medical parameters like transmission probability per contact are not directly available since surveys, usually with small sample size, give only integrated quantities like number of infected wives from men infected by transfusion after varying time elapsed since infection of the husband and after an unknown number of sexual contacts. There are nontrivial problems of parameter estimation involved; up to now there are little results from sophisticated data analyses available.

Nonetheless one has to try to use a model incorporating all relevant aspects of the epidemic dynamics in interpreting and understanding the available epidemiological data. The results from even the "simple model" presented in the last paragraph have shown that the graph structure does have a sizeable influence on the epidemic dynamics evolving on it. In addition there are questions which can be asked only to a model which simulates single individuals and single acts of infection; an important application will be mentioned below.

I would like to end by outlining the next steps we plan to take in developing and applying our model. The computer programs will be modified to simulate the situation described as the "extended model". Then, to be able to directly compare results of our model with those from standard epidemiological models, we will simulate different situations, which will be distinguished by the underlying graph structure and details of the dynamics, where these situations, after mapping them onto the coarser parameter set of the standard model, will all coincide with the parameters of the latter. This can be done with respect to other sexually transmitted diseases as well, where more data is available. Further refinements of our own modelling assumptions will be defined in collaboration with field epidemiologists. We have in mind to use the possibility to follow the fate of single individuals of our model population, using protocols, to simulate a promising prevention scenario (which has been advocated by Dietz [Di3]): as soon as an infected person comes to counseling, try to identify his past sexual contacts, include them into the counseling and give all of them advice to influence their future behaviour; the opinion is, that this kind of targeted intervention might effectively hinder transmission by lowering the growth rate of infection chaines below critical values. An important task is to bring together the purely mathematical and the simulational aspects of our work by simulating situations simple enough to still be able to derive analytical results. There is, not the least, the analysis of graphs under two aspects: to study properties of those random graphs which occur naturally in epidemiological models and to analyse the algorithms used in constructing representative graphs.

Evidently, this approach is not limited to the epidemic dynamics of sexually transmitted diseases; the general theme might be denoted as "spreading phenomena on irregular structures". You are invited to come up with nice examples besides those usually connected with graph theory like network problems. In physics, one might think of flow in amorphous inhomogeneous media. It seems to be less speculative to think of taking the other direction on this road between Mathematical Physics and Random Graph Epidemics: applying notions and methods from statistical physics and percolation theory to the analysis of the dynamics on the graph and, through it, of the graph structure.

It is a pleasure to thank Philippe Blanchard and Tyll Krüger for the joys and sorrows of our collaboration. Discussions with Prof. Dietz at various occasions helped to pave the way to a better appreciation of the mode of thinking in and the needs of epidemiology. Thanks to the people from the Centre de Physique Théorique in Marseille, especially to Daniel Testard, for organizing this workshop.

REFERENCES

[AIFO] Documentation in AIFO 3 (2), 111 (1988)

[An] Anderson, R.M., et.al.: A Preliminary Study of the Transmission Dynamics of the Human Immunodeficiency Virus (HIV), the Causative Agent of AIDS. IMA J.Math.Appl.Med.Biol. 3, 229 (1986)

[Ba] Bailey, N.T.J.: The Mathematical Theory of Infections Diseases and its Applications. London: Ch. Griffin, 1975

[BBK] Blanchard, P., Bolz, G.F., Krüger, T.: Simulation on Random Graphs of the Epidemic Dynamics of Sexually Transmitted Diseases. BiBoS preprint 291/87

[Bl] Blanchard, Ph.: A Stochastic Growth Model on Random Graphs to Understand the Dynamics of AIDS-Epidemic. To appear in: Stochastic Methods in Mathematics and Physics; Proceedings of the XXIV Karpacz Winter School of Theoretical Physics, R. Gielerak (ed.). Singapore: World Scientific, 1988

[BK1] Blanchard, Ph., Krüger, T.: Isomorphism between Epidemic Dynamical Systems on Random Graphs with Rational Transmission Probability $\gamma < 1$ and Transmission Probability One. To appear in: Stochastic Processes, Physics and Geometry; Proceedings of the 2nd Ascona Int. Conf., Lecture Notes in Physics. Heidelberg: Springer 1989

[BK2] Blanchard, Ph., Krüger, T.: Spread of AIDS-Epidemics: A Discrete Stochastic Model on Random Graphs; I. Stationary Analysis for Graphs Generated by Independent Matchings. BiBoS preprint 1988

[Bo] Bolz, G.F.: Using Random Graphs to Model the Spread of AIDS - Results from a Simple Model. in preparation

[Bs] Bollobás, B.: Random Graphs. London: Academic Press, 1985

[Di1] Dietz, K.: The Dynamics of Spread of HIV Infection in the Heterosexual Population. Preprint, Inst. of Medical Biometry, Univ. Tübingen

[Di2] Dietz, K.: On the Transmission Dynamics of HIV Math. Biosc., to appear

[Di3] Dietz, K.: private communication

[DH] Dietz, K., Hadeler, K.P.: Epidemiological Models for Sexually Transmitted Diseases. J. Math. Biol. 26, 1 (1988)

[Fr] Frauenthal, J.C.: Mathematical Modelling in Epidemiology. Heidelberg: Springer 1980

[HS] Hyman, J.M., Stanley, E.A.: Using Mathematical Models to Understand the AIDS Epidemic. Math. Biosc., to appear

[HY] Hethcote, H.W., Yorke, J.A.: Gonorrhea Transmission Dynamics and Control. (Lecture Notes in Biomathematics 56). Heidelberg: Springer, 1984

[Kn] Knox, E.G.: A Transmission Model for AIDS. Eur. J. Epidemiol. 2, 165 (1986)

[Ko] Koch, Michael G.: AIDS - vom Molekül zur Pandemie. Heidelberg: Spektrum-der-Wissenschaft-Verlags-Ges., 1987

[Si] Siegal, F.P., et al.: Severe Acquired Immunodeficiency in Male Homosexuals, Manifested by Chronic Perianal Ulcerative Herpes Simplex Lesions. N. Engl. J. Med. 305, 1439 (1981)

[WSK] Weyer, J., Schmidt, B.C., Körner, B.: Ein Mehrgruppenmodell zur Simulation der epidemischen Dynamik von AIDS, AIFO 3, 154 and 206 (1988)

SUR UNE VERSION A DEUX DIMENSIONS DU MODELE DE GINZBURG ET LANDAU POUR LA SUPRACONDUCTIVITE

A.Boutet de Monvel-Berthier, Université de Paris VI, U.A.213,
Mathématiques, 4 Place Jussieu, 75252 Paris Cedex 05.

I Description du problème.

Ginzburg et Landau ont proposé en 1950 un modèle pour la supraconductivité (état dans lequel certains matériaux, sous faible température et faible champ magnétique, acquièrent une conductivité infinie). L'état d'un supraconducteur est décrit par un champ Φ à valeurs complexes en interaction avec un champ magnétique B. $|\Phi|^2$ représente en gros la densité de présence de paires d'électrons responsables de la supraconductivité. On a $\Phi=0$ pour un conducteur normal, $|\Phi|=1$ pour un supraconducteur, et $|\Phi|\leq 1$ en général (ceci sera démontré pour le modèle que nous considérons ci-dessous).

Nous nous intéressons ici au cas d'un supraconducteur cylindrique (configuration invariante par un groupe à un paramètre de translations) dans le cas où le champ magnétique B est parallèle à l'axe. Il s'agit donc d'un problème à deux dimensions. Nous notons $\Omega\subset\mathbb{R}^2$ la section du conducteur et nous supposons que c'est un domaine borné, de bord $\Gamma=\partial\Omega$

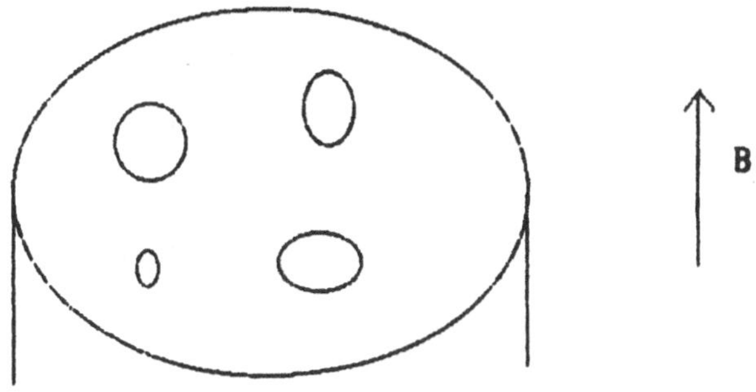

S. Albeverio et al. (eds.), Stochastics, Algebra and Analysis in Classical and Quantum Dynamics, 53–61.

Le champ magnétique B dérive d'un potentiel vecteur $A=(A_1,A_2)$, ie. $B=B_A=\partial_1 A_2-\partial_2 A_1$. On note ∇_A la connexion $\nabla_A=\nabla-iA$, et $D_A=-i\nabla_A=D-A$. On note aussi $\cdot$ la rotation d'angle $+\pi/2$ ($\cdot=\begin{pmatrix}0&-1\\1&0\end{pmatrix}$). On introduit encore la variable complexe $z=x_1+ix_2$ et on pose $\partial=\partial/\partial z=\frac{1}{2}(\partial_1-i\partial_2)$, $\bar{\partial}=\partial/\partial\bar{z}=\frac{1}{2}(\partial_1+i\partial_2)$, $\alpha=\frac{1}{2}(A_1+iA_2)$. Avec ces notations les équations du problème s'écrivent

$$(1.1)\quad \begin{aligned} &D_A^2\Phi = \tfrac{1}{2}\kappa\Phi(1-|\Phi|^2)\\ &\cdot\nabla B_A = -\operatorname{Re}(D_A\Phi.\bar{\Phi})\\ &|\Phi| = 1 \text{ sur } \partial\Omega \end{aligned}$$

où les inconnues sont Φ: $\Omega\rightarrow\mathbb{C}$ et A: $\Omega\rightarrow\mathbb{R}^2$. κ est une constante réelle qui dépend de la température; le comportement des solutions est différent selon qu'on a $\kappa>1$ (supraconducteur de type II) ou $\kappa<1$ (supraconducteur de type I). La condition limite ignore les effets de surface sur le bord Γ, et revient à considérer que le matériau est un supraconducteur parfait à la frontière. Les équations (1.1) expriment que les configurations d'équilibre sont les configurations stationnaires pour l'énergie:

$$(1.2)\ E_\kappa(A,\Phi) = \frac{1}{2}\int_\Omega [|\nabla_A\Phi|^2 + |B_A|^2 + \frac{\kappa}{4}(|\Phi|^2-1)^2]$$

Nous montrerons que

1)L'espace des configurations n'est pas connexe; ses composantes connexes correspondent aux classes d'homotopie de $\Phi|_\Gamma$

2)Les configurations extrémales sont régulières.

3)Dans le cas $\kappa=1$ nous exhibons dans chaque composante connexe les configurations qui minimisent l'énergie.

2. Espace des configurations

L'intégrale d'énergie (1.2) a un sens si les composantes de A et Φ appartiennent à l'espace de Sobolev H^1 des fonctions dont les dérivées premières sont de carré sommable dans Ω: c'est clair pour le deuxième terme de l'intégrale (1.2); et on a $\Phi \subset L^p$ pour tout p si $\Phi \in H^1$, en particulier $A\Phi$ et $(|\Phi|^2-1)^2$ sont alors intégrables, donc aussi le premier et le troisième terme. En outre si Φ est de classe H^1 la restriction $\Phi|_\Gamma$ est bien définie (au sens fonctionnel) et appartient à $H^{1/2}(\Gamma)$, ce qui permet de donner un sens à la condition limite $|\Phi|_\Gamma|=1$ Nous choisissons comme espace de configurations l'espace

$$C_1 \subset C = H^1(\Omega,\mathbb{R}^2)\times H^1(\Omega,\mathbb{C}) \text{ des couples } (A,\Phi) \text{ tels que } |\Phi|_\Gamma|=1$$

Sur cet espace de configurations opère le groupe de jauge $G=H^2(\Omega,\mathbb{R})$ des transformations unitaires, par

$$(A,\Phi)\rightarrow(A+\nabla\Lambda, e^{i\Lambda}\Phi) \text{ pour } (A,\Phi)\in C,\ \Lambda\in G$$

Les configurations physiques sont les classes d'équivalence mod. G des éléments de C_1. Les quantités qui ont une signification physique sont celles qui sont invariantes par l'action de G. Ainsi il est immédiat que l'energie (1.2) et les équations (1.1) sont invariantes par cette action (de sorte aussi que notre espace de configurations est le plus "raisonable", mais pas le plus grand possible). Sont encore invariants par l'action du groupe de jauge: le champ magnétique B_A, le module $|\Phi|$, le courant

$$(2.1)\ J(A,\Phi)=-\mathrm{Re}(D_A\Phi\,\overline{\Phi})=\mathrm{Im}(\nabla\Phi.\overline{\Phi})+A\Phi\,\overline{\Phi}$$

et l'indice d'enroulement

(2.2) $F(A,\Phi) = \frac{1}{2i\pi}\int_\Gamma \overline{\Phi}\, d\Phi = \frac{1}{2\pi}(\int_\Omega B_A - \int_\Gamma \tau.J)$

où τ est le vecteur unitaire tangent à Γ.

Aussi les degrés partiels

(2.2)bis $F_j(A,\Phi) = \int_{\Gamma_j} \overline{\Phi}\, d\Phi$

où les Γ_j sont les composantes connexes de Γ (orientées comme bord de Ω). Comme nous verrons F et les F_j sont toujours des nombres entiers; les intégrales F et F_j ont bien un sens si $\Phi \in H^{1/2}$ puisqu'on a alors $d\Phi \in H^{-1/2}$, dual de $H^{1/2}$.

<u>Proposition</u>.- <u>Les configurations</u> (A,Φ) <u>de classe</u> C^∞ (<u>ie. dont les composantes sont de classe</u> $C^\infty(\overline{\Omega})$) <u>sont denses dans</u> C_1.

<u>Corollaire</u>.- <u>Les composantes connexes de</u> C_1 <u>sont classifiées par la collection</u> $(n_0,...,n_k)$ <u>des indices partiels</u> $F_j(A,\Phi)$.

Il est clair que les $(A,\Phi) \in C^\infty(\Omega,\mathbb{R}^2) \times C^\infty(\Omega,\mathbb{C})$ sont denses dans C . Il est moins évident (à cause de la condition limite non linéaire $|\Phi|=1$ sur Γ), mais néanmoins vrai, que les $(A,\Phi) \in C^\infty \cap C_1$ sont denses dans C_1 Ceci résulte de l'observation suivante: bien qu'une fonction f de $H^{1/2}(\Gamma)$, telle que $\Phi_{|\Gamma}$, ne soit pas continue, elle est "V.M.O." ie. son oscillation moyenne sur de petits intervalles I ($\ell(I)^{-2} \iint |f(x)-f(y)|\, dx\, dy$) tend vers 0 avec la longueur $\ell(I)$ de I . De ceci on déduit que les régularisées de f sont de module voisin de 1 si $|f|=1$, ou aussi bien que le prolongement harmonique de f dans Ω est de module voisin de 1 près de Γ si $|f|=1$.Et par suite f peut être approchée dans $H^{1/2}$ par des fonctions C^∞ de module 1

Il en résulte aussi que les composantes connexes de C_1 sont les mêmes que celles de $C_1 \cap C^\infty$ et sont ouvertes dans C_1. Elles sont départagées par les classes d'homotopie de $\Phi_{|\Gamma}$. Ainsi si $\Gamma_0,\ldots,\Gamma_k$ sont les composantes connexes de Γ, les composantes connexes sont les $C^{n_0\ldots n_k}$ avec $n_j = (2i\pi)^{-1}\int_{\Gamma_j} \overline{\Phi}\, d\Phi$, degré de $\Phi_{|\Gamma_j}$ (Γ_j orienté comme bord).

3. Régularité des solutions

Nous dirons qu'une configuration physique a tel ou tel type de régularité (par exemple est de classe de Sobolev H^s) si sa classe d'équivalence contient une configuration (A,Φ) ayant ce type de regularité. Commençons par l'observation suivante: si (A,Φ) est une configuration, il existe une configuration équivalente (A',Φ'), unique (à un facteur constant près pour Φ), telle que

$$\text{(3.1)}\quad \operatorname{div} A' = 0$$
$$\nu \,.\, A' = 0 \ \text{sur}\ \Gamma$$

ν étant le champ de vecteurs normal extérieur. En effet le problème de Neumann

$$\text{(3.2)}\quad \operatorname{div} A + \operatorname{div} \nabla\Lambda = 0$$
$$\nu \,.\, (A+\nabla\Lambda) = \ \text{sur}\ \Gamma$$

a une solution $\Lambda \in H^2$ unique à une constante additive près (on a div $A \in L^2(\Omega)$, et $\nu . A \in H^{1/2}(\Gamma)$, et la condition d'intégrabilité de (3.2) est clairement satisfaite). Cette configuration a la régularité optimale, en particulier elle est de classe H^s, resp. C^∞, si la configuration de départ l'est. De cette remarque on déduit.

-si Γ est de classe C^∞ et si une configuration (A,Φ) est localement équivalente a une configuration de classe H^s ou C^∞, elle l'est globalement

- si on rajoute aux équations des configurations stationnaires (1.1)les équations (3.2), on obtient un système elliptique

$$(3.3)\quad D^2\Phi = 2A.D\Phi - A^2\Phi + \kappa/2\,(1-|\Phi|^2)\Phi$$
$$D^2A = \mathrm{Re}(\bar{\Phi}.D_A\Phi) - A|\Phi|^2$$

dont les solutions suffisamment régulières sont C^∞ (analytiques si Γ est analytique)

On a aussi le résultat suivant.

<u>Proposition</u>. - <u>Soit</u> $(A,\Phi)\in C_1$ <u>une configuration stationnaire. Alors</u>

1. <u>ou bien</u> $|\Phi|\equiv 1$, <u>ou bien</u> $|\Phi|<1$ <u>dans</u> Ω

2. <u>si de plus</u> $\kappa\leqslant 1$, <u>ou bien</u> $|B_A|\equiv(1-|\Phi|^2)/2$, <u>ou bien</u> $|B_A|<(1-|\Phi|^2)/2$ <u>dans</u> Ω.

En effet posons $w=(1-|\Phi|^2)/2 \in H_0^1(\Omega)$. Les équations (1.1) impliquent

$$(D^2+\kappa|\Phi|^2)w = |D_A\Phi|^2 \geqslant 0$$

et la première assertion résulte du principe du maximum pour w ($w>0$ ou $w\equiv 0$).En outre on a

$$(D^2+|\Phi|^2)\, B_A = i\,(\overline{D_A\Phi})(*D_A\Phi)$$

d'où on déduit

$$(D^2+|\Phi|^2)(w\pm B_A) = (1-\kappa)|\Phi|^2 + |D_A\Phi|^2 \pm i\,(\overline{D_A\Phi})\,(*D_A\Phi) \geqslant 0 \quad \text{si } \kappa\leqslant 1$$

d'où la deuxième assertion, toujours d'après le principe du maximum.

4. <u>Le cas $\kappa=1$ Configurations multi-vortex</u>

Nous faisons une étude plus détaillée des configurations minimales dans le cas $\kappa=1$

Nous notons $\Gamma_0,\dots,\Gamma_k$ les composantes connexes de Γ (Γ_0 désignant la composante extérieure, qui borde la composante non bornée de $\mathbb{C}-\Omega$), et si $n_0,\dots,n_k$ est une suite d'entiers nous notons $C^{n_0\dots n_k}$ la composante connexe de C_1 formée des configurations (A,Φ) telles que $F_j(A,\Phi) = \int_{\Gamma_j} \overline{\Phi}\, d\Phi = n_j$ Nous poserons $n = n_0+\dots+n_k$, et supposerons dans ce qui suit $n \geqslant 0$ (le cas $n \leqslant 0$ se déduit du cas considéré ci-dessous par passage au complexe conjugué). Pour $1 \leqslant j \leqslant k$ soit a_j un point de la composante de $\mathbb{C}-\Omega$ limitée par Γ_j. Soit $z_1,\dots,z_n$ une suite de n points (nopn nécessairement distincts) de Ω Posons

$$G(z) = \prod(z-z_j)\prod(z-a_j)^{-n_j}$$

C'est une fonction holomorphe dans Ω, qui ne s'annule pas sur Γ, et dont le degré sur Γ_j (orienté comme bord) est précisément n_j pour $0 \leqslant j \leqslant n$

Si maintenant $f \in H^{1/2}(\Gamma)$, $|f| = 1$, et $\deg(f_{|\Gamma_j}) = n_j$, fG^{-1} est de degré nul sur chaque Γ_j, si elle est continue son logarithme est uniforme; on montre en fait qu'on a toujours sous les hypothèses ci-dessus $f = G\, e^{i\varphi}$, avec $\varphi \in H^{1/2}(\Gamma,\mathbb{R})$ (par exemple le prolongement harmonique $\tilde{f}$ de fG^{-1} est de classe H^1, de norme voisine de 1 près de Γ, et de degré topologique nul ; son logarithme (près de Γ), qui est une primitive de $\tilde{f}^{-1} d\tilde{f}$ est alors uniforme, de classe H^1, et on en prend la restriction à Γ).

On peut réécrire ainsi l'énergie (1.2).

$$(4.1)\quad E_\kappa(A,\Phi) = 1/4\, \|(D_A \mp i \cdot D_A)\Phi\|^2 + 1/2\, \|B_A \mp w\|^2 + (\kappa-1)/2\, \|w\|^2 \pm \pi F(A,\Phi)$$

avec

$$w = 1/2\,(1-|\Phi|^2)\ ,\quad F(A,\Phi) = \text{degré (formule (2.2))}$$

On étudie maintenant plus particulièrement le cas $\kappa=1$. Dans ce cas le terme en $\|w\|^2$ disparaît et (4.1) montre que sur l'ensemble C^n des configurations de degré total n l'énergie est $\geqslant |n|$. Nous exhibons dans ce cas une famille de configurations pour lesquelles ce minimum est atteint. On se limite ici au cas $n \geqslant 0$ (l'autre cas s'en déduit par symétrie comme indiqué plus haut)

Soit G comme ci dessus (paramétrée par $n_1 \ldots n_k$, qui définissent une composante connexe de C_1, et le diviseur $z_1 + \cdots + z_n$). Cherchons A, $\Phi = G\psi$ de sorte qu'on ait

$$(4.2)\quad (D_A + i * D_A)\,\Phi = 0\ ,\qquad B_A + w = 0\,.$$

Ceci s'écrit encore

$$(4.3)\quad (\bar\partial - \alpha)\,\Phi = 0\ ,\qquad \mathrm{Re}\,\partial\alpha = w/4 \qquad \text{avec } \alpha = (iA_1 - A_2)/2$$

Comme G est holomorphe la première équation s'écrit encore

$$(4.4)\quad \alpha = \bar\partial\,\Phi/\Phi = \bar\partial\,\psi/\psi$$

Posant $a = |G|^2$, $|\psi|^2 = e^u$, $\varphi = G^{-2}|_\Gamma$ on voit que u doit vérifier l'équation

$$(4.5)\quad -\Delta u + a\,e^u = 1\ ,\qquad u|_\Gamma = \varphi$$

Cette dernière équation est variationelle coercive (parceque $a \geqslant 0$) et a une solution (unique, régulière). Dans ces conditions, u étant solution de (4.5), on peut poser $\Phi = e^{u/2}\,G$, et A est alors déterminé par (4.2) et (4.3).

Bibliographie :

[1] A.Boutet de Monvel-Berthier, V.Georgescu, R.Purice, A boundary value problem related to the Ginzburg-Landau model, à paraîre

[2] A.Boutet de Monvel-Berthier, V.Georgescu et R.Purice. Sur un problème aux limites de la théorie de Ginzburg-Landau. CRAS t.307, série I, p.55-58, 1988.

[3] L.Boutet de Monvel, O.Gabber, communication personnelle.

[4] V.I.Ginzburg, L.D.Landau. J.Exp.i Teoret.Fiziki, 20, (12) 1950.

[5] A.Jaffe, C.Taubes, Vortices and monopoles: structure of static gauge theories. Birkhäuser 1980.

DIFFERENTIAL CALCULUS AND INTEGRATION BY PARTS ON POISSON SPACE

Eric A. Carlen†
Department of Mathematics
Princeton University
Princeton, NJ 08544
USA

Etienne Pardoux‡
Mathématiques, UA 225
Université de Provence
13 331 Marseille cedex 3
France

ABSTRACT: We define a gradient operator on random variables defined on the "standard Poisson space" (the sample space of paths which have unit jumps and are constant between their jumps). An "integration by parts" formula shows that the adjoint of that operator extends the usual Poisson stochastic integral. We prove a "Malliavin calculus" type of result, which is closely related to the co-area formula of geometric measure theory.

Introduction:

The differential calculus on Wiener space (see e.g. the exposition in Watanabe [11]) is an essential tool for at least two recent developments in the theory of stochastic processes: for the "Malliavin calculus", which is a probabilistic technique for proving hypoellipticity results, and for the extension of Itô's integral and calculus to the case of antipating integrands.

Our goal in this paper is to develop an anologous calculus on Poisson space, our motivation being both to be able to prove existence of densities for functionals defined on Poisson space, and the construction of an analogue of the "Skorohod integral". More precisely, we shall work on standard Poisson space, i.e. the jump size

---

†Supported by an NSF postdoctoral fellowship.

‡The research was carried out while this author was visiting the Institute for Advanced Study, Princeton NJ, and was supported by a grant from the RCA corporation.

S. Albeverio et al. (eds.), Stochastics, Algebra and Analysis in Classical and Quantum Dynamics, 63–73.

of our Poisson process will be fixed equal to one. Note that Bichteler-Gravereaux-Jacod [1], Bismut [2] and Léandre [8] have already applied the Malliavin Calculus to Poisson driven stochastic differential equations. But their hypotheses exclude the case of fixed jump size. On the other hand, one may question the usefulness of defining a Skorohod integral with respect to Poisson process, since integrals of anticipating processes with respect a Poisson process can be defined as Stieltjes integrals. However, different integrals have different purposes. It is interesting to have an analog of the Skorohod integral, which is in particular an integral with zero expectation. We will produce such an integral here on Poisson space.

Let us now discuss our specific approach. One way of defining both the gradient operator and the divergence (i.e. Skorohod's integral) on Wiener space is to use the Wiener chaos decomposition of the set of square integrable random variables. Dermoune, Kree and Wu [5] have adapted this approach to the Poisson case. However, the "gradient" which they define in this way is not a derivation operator, and it does not seem possible to base a Malliavin-type calculus on this approach.

Rather, our gradient will be a true derivation which we introduce via variations in the intensity of the process. This approach brings into play an analog of the Girsanov theorem which plays the same role regarding variations of the intensity for the Poisson case as does the Girsanov theorem regarding translations in directions belonging to the Cameron-Martin space for the Wiener case. In contrast to the Wiener case, this approach leads to operators totally different from those found following the Poisson chaos approach.

The aim of this note is to introduce the machinery and give some preliminary results on its applications. The detailed proofs, as well as more applications will appear elsewhere.

1. Directional derivation on Poisson space.

Let Ω be the set of all maps $\omega : [0,1] \to \mathbb{N}$ such that $\omega(0) = 0$, ω is increasing and right continuous, and has finitely many jumps each of size one. Let Ω_n be the set of those maps ω with exactly n jumps. The times of these jumps completely specify such an ω and so we may identify Ω_n with $\{(t_1,\dots,t_n) \in [0,1]^n;\ 0 < t_1 < \dots < t_n < 1\}$, and topologize it accordingly. We finally topologize Ω by taking each Ω_n to be open in Ω, and denote by $\mathcal{F}$ the Borel field on Ω. Define $N_t : \Omega \to \mathbb{N}$ by $\mathbb{N}_t(\omega) = \omega(t), t \in [0,1]$. Let P denote standard Poisson measure, that is P makes $\{N_t\}$ a standard Poisson process or in other words $N_t - t$ is a martingale under P.

For every open set $\mathcal{O} \subset [0,1]$, define the σ-algebra $\mathcal{F}_{\mathcal{O}} = \sigma\{N_t - N_s, (s,t) \subset \mathcal{O}\}$. For a closed set $\mathcal{C}$, define

$$\mathcal{F}_{\mathcal{C}} = \cap_{\{\mathcal{C}\subset\mathcal{O},\mathcal{O}\text{ open}\}}\mathcal{F}_{\mathcal{O}}.$$

We shall write $\mathcal{F}_t$ for $\mathcal{F}_{[0,t]}$ and $\mathcal{F}^t$ for $\mathcal{F}_{[t,1]}$.

Let T_i denote the time of the i-th jump. That is :

$$T_i(\omega) = \begin{cases} \inf\{t; N_t(\omega) \geq i\}, & \text{if such a } t \text{ exists;} \\ 1, & \text{if } N_1(\omega) < i. \end{cases}$$

Let $\mathcal{H}$ be the subspace of $L^2(0,1)$ orthogonal to the constant functions. The elements m of $\mathcal{H}$ will be used to define changes of the intensity through time-changes. The role of the constraint $\int_0^1 m(t)\,dt = 0$ is to ensure that the change of intensity we are about to define simply shifts the times of the jump without affecting their total number. Let $m \in \mathcal{H}$. We define :

$$\tilde{m}_\varepsilon(t) = \begin{cases} -\frac{1}{3\varepsilon}, & \text{if } m(t) \le -\frac{1}{3\varepsilon}; \\ m(t), & \text{if } -\frac{1}{3\varepsilon} \le m(t) \le \frac{1}{3\varepsilon}; \\ \frac{1}{3\varepsilon}, & \text{if } m(t) \ge \frac{1}{3\varepsilon}. \end{cases}$$

and $m_\varepsilon(t) = \tilde{m}_\varepsilon(t) - \int_0^1 \tilde{m}_\varepsilon(t)\,dt$. It is easily seen that $\frac{1}{3} \le 1 + \varepsilon m_\varepsilon(t) \le \frac{5}{3}$, a technical point useful when actually carrying out the proofs which we sketch below, and that $\|m - m_\varepsilon\|_{\mathcal{H}} \to 0$, as $\varepsilon \to 0$. For the time being $m \in \mathcal{H}$ will be fixed, and we suppress reference to it in the notations that follow.

Define a reparametrization of $[0,1]$ by :

$$\tau_\varepsilon(t) = t + \varepsilon \int_0^t m_\varepsilon(s)\,ds$$

Note that $\tau_\varepsilon(0) = 0, \tau_\varepsilon(1) = 1$ and $\frac{d\tau_\varepsilon}{dt}(t) > 0, t \in [0,1]$. Next let $\mathcal{T}_\varepsilon : \Omega \to \Omega$ be the map defined by :

$$(\mathcal{T}_\varepsilon(\omega))(t) = \omega(\tau_\varepsilon(t))$$

and $P^\varepsilon = P\mathcal{T}_\varepsilon^{-1}$. Under P^ε, $\{N_t\}$ is still a Poisson process, but (E^ε denotes expectation under P^ε)

$$\begin{aligned} E^\varepsilon(N_t - N_s) =& E(N_{\tau_\varepsilon(t)} - N_{\tau_\varepsilon(s)}) \\ =& t - s + \varepsilon \int_s^t m_\varepsilon(r)\,dr \end{aligned}$$

i.e. under P^ε, $\{N_t\}$ is a Poisson process with intensity $1 + \varepsilon m_\varepsilon(t)$. Since

$$\int_0^1 \left| \frac{d\tau_\varepsilon}{dt}(t) - 1 \right|^2 dt = \varepsilon^2 \|m_\varepsilon\|^2_{\mathcal{H}} < \infty,$$

we have:

Proposition 1.1. *(Brown [4]) P^ε is absolutely continuous with respesct to P, and:*

$$\frac{dP^\varepsilon}{dP} = \prod_{i=1}^{N_1} (1 + \varepsilon m_\varepsilon(T_i))$$

□

We shall write below $\mathcal{T}_\varepsilon F$ for $F \circ \mathcal{T}_\varepsilon$, if $F \in L^2(\Omega)$. Let us define the set:

$$\mathbb{D}^0_m = \{F \in L^2(\Omega); L^2(\Omega) - \lim_{\varepsilon \to 0} \frac{1}{\varepsilon}(\mathcal{T}_\varepsilon F - F) \,\text{exists}\}.$$

For $F \in \mathbb{D}^0_m$, we define $D_m F$ as the above limit. Let us consider some basic examples.

Proposition 1.2. *Each T_i belongs to each $\mathbb{D}^0_m$ and*

$$D_m T_i = -\int_0^{T_i} m(t)\,dt$$

Proof:

$$|T_\varepsilon(T_i) - T_i + \varepsilon \int_0^{T_i} m_\varepsilon(t)\,dt| \le c\varepsilon^2$$

where c depends only on $\|m\|_{\mathcal{H}}$. □

Next, for $h \in \mathcal{H} \cap C([0,1])$, set

$$\begin{aligned}\delta(h) &= \sum_{i=1}^{N_1} h(T_i)\\ &= \int_0^1 h(t)\,dN_t\\ &= \int_0^1 h(t)\,d(N_t - t)\end{aligned}$$

with the convention that the sum over no jumps is zero. The last equality follows since $\int_0^1 h(t)\,dt = 0$; $\delta(h)$ is the *compensated* integral of h against dN.

Since

$$T_\varepsilon(\delta(h)) - \delta(h) = \sum_{i=1}^{N_1} (h(T_\varepsilon(T_i)) - h(T_i)),$$

it is easy to see, using the mean value theorem and dominated convergence, that whenever $h \in \mathcal{H} \cap C^1([0,1])$, $\delta(h) \in \mathbb{D}^0_m$ for any $m \in \mathcal{H}$ and

$$D_m\delta(h) = -\sum_{i=1}^{N_1} h'(T_i) \int_0^{T_i} m(t)\,dt.$$

Notice that this is not deterministic, in contrast with the analogous quantity on Wiener space. Notice though, that

$$E\, D_m\delta(h) = < m, h >_{\mathcal{H}}.$$

To proceed we introduce the class of simple functions $\mathcal{S}$. $F \in \mathcal{S}$ in case F can be written as

$$F = f(T_1, \ldots, T_n)$$

for some n and some C^1 function f on the simplex $\Delta_n = \{(t_1, \ldots, t_n) | 0 < t_1 < t_2 < \cdots < t_n < 1\}$ which has a continuous extension together with its first derivatives to the closure of Δ_n.

Theorem 1.3. *$\mathcal{S}$ is dense in $L^2(\Omega)$. Moreover, $\mathcal{S} \subset I\!D^0_m$, for any $m \in \mathcal{H}$ and*

$$D_m f(T_1, \ldots, T_n) = -\sum_{j=1}^{n} \frac{\partial f}{\partial t_j}(T_1, \ldots, T_n) \int_0^{T_j} m(t)\, dt$$

Proof: This follows easily from Proposition 1.2. □

Theorem 1.4. *For all F, G in $\mathcal{S}$ and m in $\mathcal{H}$,*

$$E(D_m F) = E(\delta(m) F)$$

$$D_m(FG) = (D_m F)G + F(D_m G)$$

Proof: We only sketch the proof of the first statement. The proof of the second one follows from the formula at the end of the previous proof.

$$\begin{aligned} E(D_m F) &= \lim_{\varepsilon \to 0} \frac{1}{\varepsilon} E(T_\varepsilon F - F) \\ &= \lim_{\varepsilon \to 0} \frac{1}{\varepsilon} E\left[\left(\frac{dP^\varepsilon}{dP} - 1\right) F\right] \\ &= \lim_{\varepsilon \to 0} E[F \sum_{i=1}^{N_1} m_\varepsilon(T_i)] \\ &= E[F\delta(m)] \end{aligned}$$

□

Theorem 1.5. *For any $m \in \mathcal{H}$, D_m is an unbounded closable densely defined operator from $L^2(\Omega)$ into itself.*

Proof: The "densely defined" statement follows from the fact that $\mathcal{S}$ is dense in $L^2(\Omega)$. The "closable" statement follows easily from the following formula, which is valid for $F, G \in \mathcal{S}$, $m \in \mathcal{H}$ and follows from Theorem 1.4:

$$E(G D_m F) = E[GF\delta(m)] - E[F D_m G]$$

□

We shall identify D_m with its closed extension, and denote by $I\!D^{1,m}$ its domain. Clearly each D_m is a derivation. While D_{m_1} and D_{m_2} do not in general commute, $D_{m_1} D_{m_2} - D_{m_2} D_{m_1}$ is another such operator, namely: $D_{S(m_1, m_2)}$ where

$$S(m_1, m_2) = m_1'(t) \int_0^t m_2(s)\, ds - m_2'(t) \int_0^t m_1(s)\, ds \in \mathcal{H}.$$

2. The gradient operator

We want next to define an operator D from $L^2(\Omega)$ into $\mathcal{H} \otimes L^2(\Omega)$ which should have the poperty that any F in its domain should belong to all $\mathbb{D}^{1,m}$, and should satisfy:

$$\int_0^1 D_t F m(t)\, dt = D_m F$$

It is then natural to define D on $\mathcal{S}$ as follows. For $F \in \mathcal{S}$, $F = f(T_1, \ldots, T_n)$,

$$D_t F = -\sum_{j=1}^{n} \frac{\partial f}{\partial t_j}(T_1, \ldots, T_n) \left(1_{[0,T_j]}(t) - T_j\right).$$

Note that $1_{[0,T_j]} - T_j$ is the orthogonal projection in $L^2([0,1])$ of $1_{[0,T_j]}$ onto $\mathcal{H}$. An argument similar to that of Theorem 1.4 yields:

Theorem 2.1. *D is an unbounded closable densely defined operator from $L^2(\Omega)$ into $\mathcal{H} \otimes L^2(\Omega)$.* □

Again, we identify D with its closed extension, and denote its domain by $\mathbb{D}^{1,2}$.

Note that for any $F \in \mathbb{D}^{1,2}$, the process $\{D_t F, t \in [0,1]\}$ is constant on each interval (T_i, T_{i+1}), since it is the case for $F \in \mathcal{S}$. It is easily seen that if $F \in L^2(\Omega)$ is $\sigma(N_1)$ measurable, then $F \in \mathbb{D}^{1,2}$ and $D_t F \equiv 0$. Moreover, for $i \geq 1$, $T_i \in \mathbb{D}^{1,2}$ and $D_t T_i = T_i - 1_{[0,T_i]}(t)$.

Again, it is clear that D is a derivation, i.e. $D_t(FG) = G D_t F + F D_t G$. Note that if $F \in \mathcal{F}_{[t_1,t_2]}$, $D_t F$ is a.s. constant on $(0, t_1)$ and $(t_2, 1)$.

3. The divergence operator

Let $\delta : \mathcal{H} \otimes L^2(\Omega) \to L^2(\Omega)$ denote the adjoint of D. Then its domain, Domδ, is the set of $u \in \mathcal{H} \otimes L^2(\Omega)$ which are such that there exists $c > 0$ with :

$$|E \int_0^1 D_t F u_t\, dt| \leq c \|F\|_2, \ \forall F \in \mathbb{D}^{1,2}$$

For u $\in$Domδ, we define $\delta(u)$ as the unique element of $L^2(\Omega)$ (whose existence follows from Riesz's representation theorem) which satisfies:

$$E(\delta(u)F) = E \int_0^1 u_t D_t F\, dt, \ \forall F \in \mathbb{D}^{1,2}$$

It follows from the properties of D that δ is a closed densely defined operator. A basic fact about the operator δ is that it generalizes the Itô integral.

Theorem 3.1. *Let $u \in \mathcal{H} \otimes L^2(\Omega)$ be predictable. Then $u \in Dom\delta$ and*

$$\delta(u) = \int_0^1 u_t \, dN_t$$

where the above is the compensated Poisson stochastic integral, which is just the ordinary Stieltjes integral if $\{u_t\}$ is left continuous. □

The proof is based on an approximation argument and the closedness of δ. Notice in particular whan u is deterministic, i.e. $u \in \mathcal{H}$, $u \in Dom\delta$ and $\delta(u) = \sum_{i=1}^{N_1} u(T_i)$, which justifies the notation of section 1. An important property of the operator δ, which follows from the fact that D is a derivation, is :

Theorem 3.2. *Let $F \in \mathbb{D}^{1,2}$ and $X \in Dom\delta$ be such that :*

$$F\delta(X) - \int_0^1 D_t F X_t \, dt \in L^2(\Omega)$$

Then $FX \in Dom\delta$ and :

$$\delta(FX) = F\delta(X) - \int_0^1 D_t F X_t \, dt$$

Proof: Let $G \in \mathcal{S}$. Then :

$$\begin{aligned} E\int_0^1 FX_t D_t G \, dt =& E\int_0^1 D_t(GF)X_t \, dt - -E\int_0^1 GD_tFX_t \, dt \\ =& E[G(F\delta(X) - \int_0^1 D_tFX_t \, dt)] \end{aligned}$$

□

To further describe the domain of δ, we introduce a class $\tilde{\mathcal{S}}$ of simple processes: we say that $u \in L^2(\Omega) \otimes \mathcal{H}$ belongs to $\tilde{\mathcal{S}}$ in case u has the form

$$u_t = u(t, T_1, \ldots, T_n)$$

where $u(\cdot, T_1, \ldots, T_n)$ is continuous for each $T_1, \ldots, T_n$, each $u_t \in \mathcal{S}$, and finally

$$\sup_{0 \le t \le 1} |u_t| \in \cap_{p>1} L^p(\Omega).$$

Theorem 3.3. *$\tilde{\mathcal{S}} \subset Dom\delta$ and for $u \in \tilde{\mathcal{S}}$,*

$$\delta(u) = \int_0^1 u_t \, dN_t - \int_0^1 D_t u_t \, dt.$$

Proof : For $F \in \mathcal{S}$, $u \in \tilde{\mathcal{S}}$, compute $E\left(\int_0^1 D_tFu_t\,dt\right)$ directly in terms of integrals on Δ_n. One integrates by parts on Δ_n, collects the boundary terms into a telescoping sum, and obtains the above result. □

Using the closure of δ and an approximation argument, we can extend this formula, with appropriate minor modifications , to a large class of discontinuous u's; in particular processes u that jump with N_t.

Note that

$$-\int_0^1 D_tu_t\,dt = \sum_{j=1}^{\infty}\int_0^{T_j}\frac{\partial u_t}{\partial T_j}\,dt$$

and if u is adapted, $\frac{\partial u_t}{\partial T_j} = 0$ for all $t < T_j$. Therefore $\int_0^1 D_tu_t\,dt = 0$ for all adapted u, and Theorems 3.1 and 3.3 are in agreement.

If one expected a complete analogy with the calculus on Wiener space, one might conjecture, based on a result of M. Kree, that $\mathbb{D}^{1,2}\otimes\mathcal{H}\subset Dom\delta$. This is not the case. In fact:

Example 3.4. *For any bounded function g on $\mathbb{R}_+$ define $F = g(N_1)$. Clearly $F \in \mathbb{D}^{1,2}$, $D_tF = 0$, $\|F\|_{\mathbb{D}^{1,2}} = \|F\|_{L^2(\Omega)}$. Define $h \in \mathcal{H}$ by $h(t) = 1$ for $t \le \frac{1}{2}$, and $h(t) = -1$ for $t > \frac{1}{2}$. Define $u_t = Fh(t)$. Clearly $u \in \mathbb{D}^{1,2}\otimes\mathcal{H}$ and $\|u\|_{\mathbb{D}^{1,2}\otimes\mathcal{H}} = \|F\|_{L^2(\Omega)}$. Moreover, it is clear by Theorem 3.2 that $u \in Dom\delta$ whenever $F\delta(h) \in L^2(\Omega)$ in which case $\delta(u) = F\delta(h)$. But an elementary computation yields*

$$E[F^2\delta(h)^2|N_1 = n] = g^2(n)n,\; E(\delta(u)^2) = E(N_1F^2).$$

Thus there is no constant c so that $\|\delta(u)\|_{L^2(\Omega)} \le c\|u\|_{\mathbb{D}^{1,2}\otimes\mathcal{H}}$.

□

4. The Malliavin Calculus

The aim of this section is to show how one can infer the existence of a density from the non-degeneracy of the Malliavin covariance matrix, and deduce from that fact a simple result for Poisson-driven stochastic differential equations. We shall restrict ourselves for simplicity to one-dimensional random variables.

First note that a random variable F on Poisson space cannot have a density, nor can the Malliavin variance $\int_0^1 |D_tF|^2\,dt$ be a.s. strictly positive. Indeed, the law of F has a point mass at $F(0)$, and $\int_0^1 |D_tF|^2\,dt = 0$ on $\{N_1 = 0\}$. Therefore, we shall rather give conditions under which $(1_{\{N_1\ge 1\}}P)F^{-1}$ has a density. The proof of the following result follows a proof in Nualart-Zakai [9] in the Wiener case :

Theorem 4.1. *Let $F \in \mathbb{D}^{1,2}$ and $B \subset \mathbb{N} - \{0\}$ be such that $\int_0^1 |D_tF|^2\,dt > 0$ a.s. on $A = \{N_1 \in B\}$. Then $(1_AP)F^{-1}$, the image by F of the restriction of P to A, is absolutely continuous with respect to Lebesgue measure on $\mathbb{R}$.*

Proof: We can without loss of generality assume that F takes values in $[-1, 1]$. We have to show that for any measurable function $g : [-1, 1] \to [0, 1]$ such that

$\int_{-1}^{1} g(x)\,dx = 0$, $E[1_A g(F)] = 0$. Let $\{g^n, n \in I\!N\} \subset C_c^1(I\!R)$ be such that $g^n(x) \to g(x)\,dx + dQ(x)$ a.e., where $Q = PF^{-1}$, and $|g^n(x)| \leq 1$, $\forall x \in I\!R$, $n \in I\!N$. Define

$$\Psi^n(x) = \int_{-\infty}^{y} g^n(x)\,dx; \; \Psi(x) = \int_{-1}^{y} g(x)\,dx$$

Then $\Psi^n(F) \in I\!D^{1,2}$ and:

$$D_t[\Psi^n(F)] = g^n(F) D_t F.$$

Moreover, $\Psi^n(F) \to \Psi(F)$ in $L^2(\Omega)$, where

$$\Psi(F) = 0 \text{ a.s.}$$

and $D\Psi^n(F) \to g(F)DF$ in $L^2(\Omega \times (0,1))$. Therefore, $g(F)DF = 0$. On A, $\int_0^1 |D_t F|^2\,dt > 0$ a.s., so that $g(F) = 0$ a.s. □

Let us now apply that result to the solution of a stochastic differential equation driven by a Poisson process. In doing so, we shall see that Theorem 4.1 is in fact a simple consequence of the co-area formula of geometric measure theory (see Federer [6]). Note that the connection between Malliavin's calculus on Wiener space and the co-area formula has been pointed out and exploited by Bouleau-Hirsch [3]. Let f and g be smooth functions from $I\!R$ into $I\!R$. We suppose moreover that for some $C > 0$:

$$|f(x)| \leq C(1 + |x|), \; \forall x \in I\!R.$$

Let $\{X_t, t \geq 0\}$ denote the solution of :

$$\begin{cases} dX_t = f(X_t)dt + g(X_{t^-})dN_t \\ X_0 = x \end{cases} \tag{4.1}$$

We want to apply Theorem 4.1 to $F = X_1$. Let $\{\Phi_t(x), t \geq 0\}$ denote the flow defined by :

$$\begin{aligned} \frac{d}{dt}\Phi_t(x) &= f \circ \Phi_t(x), \\ \Phi_0(x) &= x \end{aligned}$$

On the set $\{N_1 = 0\}$,

$$X_1 = \Phi_1(x).$$

On the set $\{N_1 = 1\}$,

$$X_1 = \Phi_{1-T_1}(\Phi_{T_1}(x) + g(\Phi_{T_1}(x)))$$

and so on. Let us compute $D_t X_1$ on the set $\{N_1 = 1\}$.

$$\begin{aligned} D_t X_1 = \exp\Big(\int_{T_1}^{1} f'(X_s)\,ds\Big)[(1 + g'(\Phi_{T_1}(x)))f(\Phi_{T_1}(x)) \\ - f(\Phi_{T_1}(x) + g(\Phi_{T_1}(x))))] D_t T_1 \end{aligned}$$

Consequently,

$$\int_0^1 |D_t X_1|^2\,dt = exp(2\int_{T_1}^1 f'(X_s)ds)[(1+g'(\Phi_{T_1}(x)))f(\Phi_{T_1}(x)) \\ -f(\Phi_{T_1}(x)+g(\Phi_{T_1}(x)))]^2(1-T_1)T_1$$

Note that when $f(0)=0$ and $g(x)=-x$, this vanishes. The solution jumps to the origin at the first jump and stays there. However a sufficient condition for $\int_0^1 |D_t X_1|^2\,dt > 0$ a.s. on $\{N_1=0\}$ is then that the following function does not vanish :

$$\begin{aligned} h(y) =& g'(y)f(y)+f(y)-f(y+g(y)) \\ =& g'(y)f(y)-f'(y)g(y)-\frac{1}{2}f''(z_y)g^2(y) \end{aligned}$$

Therefore, we have :

Proposition 4.2. *let $\{X_t, t\in[0,1]\}$ denote the solution of equation (4.1). A sufficient condition for the measure $(1_{\{N_1=1\}}P)X_1^{-1}$ to be absolutely continuous with respect to one dimensional Lebesgue measure is that :*

$$|W(g,f)(x)| > \frac{1}{2}\|f''\|_\infty \|g\|_\infty^2,\ x\in I\!R \tag{4.2}$$

where $W(g,f)(x)=g'(x)f(x)-f'(x)g(x)$ is the Wronskian of g and f. □

Note that there exists a smooth function $\varphi:(0,1)\to I\!R$ s.t. on the set $\{N_1=1\}$, $X_1=\varphi(T_1)$. Condition (4.2) implies that φ' does not vanish, which implies the conclusion of Proposition 4.2 by a very elementary argument. More generally, on the set $\{N_1=n\}$,

$$X_1=\varphi(T_1,\ldots,T_n)$$

and it follows from the co-area formula that the absolute continuity of X_1 is implied by :

$$\sum_{i=1}^n |\frac{\partial\varphi}{\partial t_i}(T_1,\ldots,T_n)|^2 > 0 \text{ a.s.}$$

The same result follows from our "Malliavin Calculus" approach. For simplicity, let us consider the case $n=2$. Then :

$$\begin{aligned} \int_0^1 |D_t X_1|^2\,dt =& \frac{\partial\varphi}{\partial t_1}(T_1,T_2))^2 T_1(1-T_1) \\ &+2\frac{\partial\varphi}{\partial t_1}(T_1,T_2)\frac{\partial\varphi}{\partial t_2}(T_1,T_2)T_1(1-T_2) \\ &+(\frac{\partial\varphi}{\partial t_2}(T_1,T_2))^2 T_2(1-T_2) \\ \geq& \inf\left((\frac{\partial\varphi}{\partial t_1}(T_1,T_2))^2\frac{T_1}{T_2}(T_2-T_1),(\frac{\partial\varphi}{\partial t_2}(T_1,T_2))^2\frac{1-T_2}{1-T_1}(T_2-T_1)\right) \end{aligned}$$

Note that on $\{N_1 = n\}$,

$$X_1 = \Phi_{1-T_n}(\Phi_{T_n}(z(T_1, \ldots, T_{n-1})) + g(\Phi_{T_n}(z(T_1, \ldots, T_{n-1}))).$$

Therefore, condition (4.2) implies that $\frac{\partial \varphi}{\partial t_n}$ does not vanish, and we obtain:

Theorem 4.3. *Under condition (4.2),* $(1_{\{N_1 \geq 1\}} P) X_1^{-1}$ *is absolutely continuous with respect to Lebesgue measure on* $\mathbb{R}$. □

Bibliography

[1] K. Bichteler, J.B.Gravereaux, J. Jacod *Malliavin calculus for processes with jumps*, Stochastics Monographs **2** Gordon & Breach (1987).

[2] J.M. Bismut *Calcul des variations stochastique et processus de saut*, Zeitschrift für Wahrschein. **63**, 147-235 (1983).

[3] N. Bouleau, F. Hisch *Propriétés d'absolue continuité dans les espaces de Dirichlet et applications aux équations différentielles stochastiques*, in Séminaire de Probabilités XX, Lecture Notes in Mathematics **1204**, 131-161 (1986).

[4] M. Brown *Discrimination of Poisson processes*, Ann. Math. Stat. **42**, 773-776 (1971).

[5] A. Dermoune, P. Krée, L. Wu *Calcul stochastique non adapté par rapport à la mesure de Poisson*, in Séminaire de Probabilité XXII, to appear.

[6] H. Federer *Geometric measure theory*, Springer Verlag (1969).

[7] K. Itô *Spectral type of the shift transformations of differential processes with stationary increments*, Trans. Amer. Math. Soc. **81**, 253-263 (1956).

[8] R. Léandre *Régularité des processus de saut dégénérés* Ann. Inst. H. Poincaré **21**, 125-146 (1985).

[9] D. Nualart, M. Zakai *The partial Malliavin calculus*, Preprint.

[10] I. Segal *Tensor algebras over Hilbert space I*, Trans. Amer. Math. Soc. **81**, 106-134 (1956).

[11] S. Watanabe *Stochastic differential equations and Malliavin calculus*, Tata Institute, Springer Verlag (1984)

INTRODUCTION TO ENTIRE CYCLIC COHOMOLOGY (OF $\mathbb{Z}/2$-GRADED BANACH ALGEBRAS)

Daniel KASTLER
Centre de Physique Théorique*
CNRS - Luminy, Case 907
F-13288, MARSEILLE CEDEX 09 (FRANCE)

§ 0 Foreword

§ 1 The differential envelope $\Omega(A)$ of a $\mathbb{Z}/2$-graded complex algebra A. The calculus within $\Omega(A)$.

§ 2 The entire cyclic cohomology of a $\mathbb{Z}/2$-graded Banach algebra A.

§ 3 The normalized entire cocycles as representing entire cyclic cohomology.

§ 4 The normalized entire cocycles of A as 1-paratraces of $\Omega(A)$.

§ 5 The Cuntz envelope QA and Zekri algebra $\mathcal{E}A$ of $\mathbb{Z}/2$-graded complex algebra A.

§ 6 The even (resp. odd) entire cyclic cocycles as odd traces of $\mathcal{E}A$ (resp. QA).

§ 7 The characters of θ-summable Fredholm modules.

Appendix A. A sketch of ordinary cyclic cohomology of $\mathbb{Z}/2$-graded complex algebras.

Appendix B. The Cuntz deformation $(Q_t\Omega,\tau,d)$ and Zekri algebra $(\mathcal{E}_t\Omega,\hat{\tau},\tau)$ of a $\mathbb{Z}/2$-graded differential algebra (Ω,θ,d).

Appendix C. Odd (resp. even) $t^2/2$-paratraces of Ω as odd traces of $Q_t\Omega$(resp. $\mathcal{E}_t\Omega$).

Appendix D. The basic maps relating Ω and $Q_t\Omega$.

Appendix E An alternative form of the character.

Appendix F The character of graded KMS - functionals

* Laboratoire Propre n. 7061, Centre National de la Recherche Scientifique
* et Université d'Aix-Marseille II

S. Albeverio et al. (eds.), Stochastics, Algebra and Analysis in Classical and Quantum Dynamics, 75–152.

§ 0 FOREWORD

By now we have compelling evidence from various horizons of mathematics (group algebras, algebras of foliations, cristalline cohomology, etc.) that the basic notion of cohomology for non-commutative algebras is Alain Connes' **cyclic cohomology** [0]. We are therefore led to expect crucial applications in physics, a hope which begins to become substanciated [3], [4], [5], [21], [22]. However, in physical applications, the concrete objects through which a cohomology is realized (e.g. the differential forms in the case of De Rham cohomology) are of greater importance than the abstract cohomology itself. In the case considered here of a non-commutative algebra A (corresponding to the quantum case in physics) the cyclic cocycles of A concretely occur as **characters** of the **Fredholm A-modules** - the latter playing the role of the elliptic operators of the classical case. Now, in cyclic cohomology as originally considered [0], one deals with cocycles a priori confined in dimension, the relevant Fredholm modules being then p-**summable**, a notion which (as noticed by Connes in [2]) only accomodates (in the test area of locally compact groups) the convolution algebras of "groups of slow growth". In order to widen the range of applicability of his theory, Connes was thus led to propose the notion of **θ-summable Fredholm module** characterized by a less stringent (physically more appealing) summability condition (of a "heat equation" or 'high temperature" flavour). These θ-summable Fredholm modules now yield cocycles of a new type of cohomology (substitute of the previous (periodic) cyclic cohomology) called by Connes **entire cyclic cohomology**, the name alluding to a growth condition of the entire function type[0] .

This report introduces to entire cyclic cohomology [1], presented as extended to $\mathbb{Z}/2$-graded Banach algebras [6] (a natural mathematical exercise motivated by physics, where "superalgebras" abound through the occurence of fermionic degrees of freedom). Accessorily we offer in an Appendix an introduction to the former cyclic cohomology [0], whose technical mathematical apparatus was used by Alain Connes to develop entire cyclic cohomology.

---

[0] this condition is responsible for the non-triviality of the cohomology (without it one has an acyclic complex)

In a previous presentation of the $\mathbb{Z}/2$-graded version of usual cyclic cohomology [7] [8], we had found it advantageous to handle the cyclic cocycles of a complex algebra A (for that matter, its Hochschild cocyles) as linear forms of the **differential envelope** $\Omega(A)$ of A (rather than traditionally as multilinear forms on A), thereby using as a tool our "calculus within the differential envelope" (cf. Chapter 2 of [7], sketched in Appendix A of [6]). This "global approach" is now particularly suited for handling the new entire cyclic cocycles "delocalized in dimension".

This report is organized as follows : we begin in Section 1 with a description of the differential envelope and its "calculus". Section 2 presents the **entire cyclic cochains, cocyles,** and **cohomology groups** (the reader interested in the relationship with usual cyclic cohomology will find a sketch of the latter in Appendix A). In section 3 we present the (subset of) **normalized cyclic cocyles**, a special kind of cocycles which, whilst representative of entire cyclic cohomology, have the virtue, as explained in Section 4, to be realized as the **paratraces** of $\Omega(A)$ (= the linear forms of $\Omega(A)$ vanishing on **parabrackets**). Section 6 then describes a (still more useful) aspect of the normalized cocycles, their description as traces (with appropriate symmetry) of the **Zekri** or **Cuntz algebra** of A (the latter described in Section 5). In fact one naturally defines the **Cuntz modification** $Q_t\Omega$ and the accompanying **Zekri algebra** $\mathcal{E}_t\Omega$ for an arbitrary $\mathbb{Z}/2$-graded differential algebra Ω : the odd (resp. even) $t^2/2$-paratraces of Ω are then respectively converted into odd traces of $Q_t\Omega$ (τ-invariant odd traces of $\mathcal{E}_t\Omega$) : this general structure is described in Appendices B and C, supplemented by Appendix D describing canonical mappings between $\Omega^\pm$ and $Q_t\,\Omega^\pm$ which turn $t^2/2$ paratraces into linear combinations of (modified) commutators. Our last section 7 attempts to motivate Alain Connes' (somewhat involved) construction of the Chern character of a θ-summable Fredholm module (this in the trivially graded case - the $\mathbb{Z}/2$-graded generalization still awaits investigation). Appendix E describes in its proper $\mathbb{Z}/2$-graded context a form of the Chern character due to Jaffee, Lesniewski and Osterwalder [21]. Appendix F shows that *"graded KMS functionals"* of $\mathbb{Z}/2$ graded algebras with *"supersymmetric"* modular groups generate cyclic cocycles (irrespective of parity) [22].

As a conclusion of this introduction, let us note the fact that the original cyclic cohomology and the new entire cyclic cohomology share the feature of having, amongst their cocycles, a subset of "normalized" cocycles interpretable as specific linear forms of the differential envelope, resp. its Cuntz or Zekri modification. Historically however, whilst in original cyclic cohomology the normalized (=cyclic Hochschild) cocycles were the first to appear, a subsequent "enlarging of the the playground" leading to the general cocyles (cf. [A.3] below), the converse happened for the entire cyclic cohomology : here the general cocycles naturally appear first, the normalized cocycles being then introduced for the sake of their "geometrical" interpretation - with the previous situation of usual cyclic cohomology serving as a guide.

§ 1 THE DIFFERENTIAL ENVELOPE $\Omega(A)$ OF A $\mathbb{Z}/2$-GRADED COMPLEX ALGEBRA A. THE CALCULUS WITHIN $\Omega(A)$.

[1.1] Let A be a $\mathbb{Z}/2$-graded complex algebra (i.e. one has $A = A^0 \oplus A^1$, with $A^i A^j \subset A^{i+j}$, $i,j = \mathbb{Z}/2$; we write $\partial a = 0 \bmod 2$ (resp. 1 mod. 2) for $a \in A^0$ (resp.A^1)). The concept of **differential envelope** of A arises from the attempt to formalize the notion of "differentials" da of elements $a\in A$. We want to construct a $\mathbb{Z}/2$-graded algebra $\Omega(A)$ generated by the $a\in A$ and their "differentials" da , so as to have d a **differential of** $\Omega(A)$, i.e. a graded derivation with vanishing square : to this aim we naturally construct $\Omega(A)$ via symbols

$$(1.1) \qquad \begin{cases} a\in A \\ da\ ,\ a\in A \end{cases}$$

and relations[1]

$$(1,2) \qquad \begin{cases} \lambda\cdot a \dot{+} \mu\cdot b \dot{-} (\lambda a+\mu b)=0 \\ a\cdot b \dot{-} (ab)=0 \\ \lambda\cdot da \dot{+} \mu\cdot db \dot{-} d(\lambda a+\mu b)=0 \\ da\cdot b+(-1)^{\partial a}\, a\cdot db \dot{-} d(ab)=0 \end{cases} , \quad \begin{cases} a\in A^0\cup A^1 \\ b\in A \\ \lambda,\mu\in \mathbb{C} \end{cases}$$

(The operations written with a • are the "formal" ones within the free algebra-ordinary notation referring to operations within A).

Clearly, by reordering any "word" with letters (1,1) by means of the last relation (1,2) (so as to have all symbols da standing to the right of the symbols a, which then conglomerate

---

[1]In other terms $\Omega(A)$ is by definition the quotient of the free algebra over $\mathbb{C}$ generated by the a and da, $a\in A$, through the ideal generated by the expressions on the left side of the relations (1,2). Clearly, the first two relations (1,2) aim at having A a subalgebra of $\Omega(A)$; and the two last ones at making d a graded derivation.

by the second relation (1,2)), we see that $\Omega(A)$ is linearly generated by symbols of the type

$$(1,3)\qquad \begin{cases} a_0 da_1 \dots\dots da_n \\ \\ da_1 \dots da_n \end{cases}, \quad a_0, a_1, \dots, a_n \in A, n \in N,$$

or, more economically (adding formally [2] a unit $\tilde{1}$ to A with the ensuing augmentation $\tilde{\Omega}(A) = \mathbb{C}\ \tilde{1} + \Omega(A)$ of $\Omega(A)$), by symbols

$$(1,4)\qquad a_0 da_1 \dots da_n \begin{cases} a_0 \in \tilde{A} = \mathbb{C}\ \tilde{1} \oplus A \\ \\ a_1, \dots, a_n \in A \end{cases}, \quad n \in N$$

[**1.2**] We are thus led to the following constructive definition of $\tilde{\Omega}(A)$: *the latter is built as the vector space*

$$(1,5)\qquad \tilde{\Omega}(A) = \bigoplus_{n \in N} \tilde{\Omega}(A)^n ,$$

where [3]

$$(1,6)\qquad \begin{cases} \tilde{\Omega}(A)^\circ = \tilde{A} = \mathbb{C}\ \tilde{1} \oplus A \\ \tilde{\Omega}(A)^n \cong \tilde{A} \otimes A^{\otimes n} = \{a_0 da_1 \dots da_n\ ;\ a_0 \in \tilde{A}, a_1, \dots a_n \in A\},\ n \geq 1 \end{cases},$$

and endowed with an associative bilinear product determined by the rule [4] .

---

[2] even though A might already possess a unit e (the latter then becomes the generating idempotent of the ideal $0 \oplus A$ in the augmented algebra $\tilde{A} = \mathbb{C}\ \tilde{1} \oplus A$).

[3] The free construction via (1,1), (1,2) makes it intuitive that we have a linear isomorphism : $a_0 da_1 \dots da_n \leftrightarrow a_0 \otimes a_1 \otimes \dots a_n$, $a_0 \in \tilde{A}$, $a_1, \dots, a_n \in A$ (for proofs see [7] Chapter 1). Note that $\tilde{\Omega}^\circ = \tilde{A}$, $\tilde{\Omega}(A)^n = \Omega(A)^n$.

[4] This rule heuristically proceeds from stepwise reordering via the last relation (1,2) - and implies an obvious definition of the product of $\tilde{\Omega}(A)$ (cf. [7], chapter 1).

$$(1,7)\qquad (a_0 da_1 \dots da_n)a_{n+1} = (-1)^{n+\sum_{k=1}^{n}\partial a_k} a_0\, a_1\, da_2 \dots da_{n+1},$$

$$+\sum_{j=1}^{n}(-1)^{n+j+\sum_{k=j+1}^{n}\partial a_k} a_0 da_1 \dots d(a_j\, a_{j+1}) \dots da_{n+1}$$

$a_0 \in \tilde{A}$, $a_1, \dots a_n \in A$ of resp. grades ∂a_0, $da_1, \dots, \partial a_n$; $a_{n+1} \in A$

It is intuitive that the above constructive definition of $\Omega(A)$ yields a $\mathbb{Z}/2$-graded complex algebra with grading[5]

$$(1,8)\qquad \partial(a_0 da_1 \dots da_n) = n + \sum_{k=0}^{n} \partial a_k \; ; \quad \begin{cases} a_0 \in \tilde{A}^0 \cup \tilde{A}^1 \\ a_1, \dots, a_n \in A^0 \cup A^1 \end{cases}$$

(formal proofs using recursion are easy to construct). It is also intuitive that the constructive definition yields back the algebra defined via the symbols and relations (1,1), (1,2). A formal proof of this relates to the following

[1.3] Universal property of $\Omega(A)$

To any complex algebra B, *and pair of linear maps* , $\varphi, \delta : A \to B$ *fufilling*

$$(1,9)\qquad \begin{cases} \varphi(ab) = \varphi(a)\, \varphi(b) \\ \delta(ab) = \delta(a)\varphi(b) + (-1)^{\partial a}\varphi(a)\delta(b) \end{cases} , \quad \begin{cases} a \in A^0 \cup A^1 \\ b \in A \end{cases}$$

there is a unique homomorphism $\theta : \Omega(A) \to B$ *of complex algebras making the following diagram commutative*

[5] We endow $\tilde{A}$ with the natural grading $\tilde{A}^\circ = \mathbb{C}\tilde{1} \oplus A^\circ$, $\tilde{A}^1 = 0 \oplus A^1$.

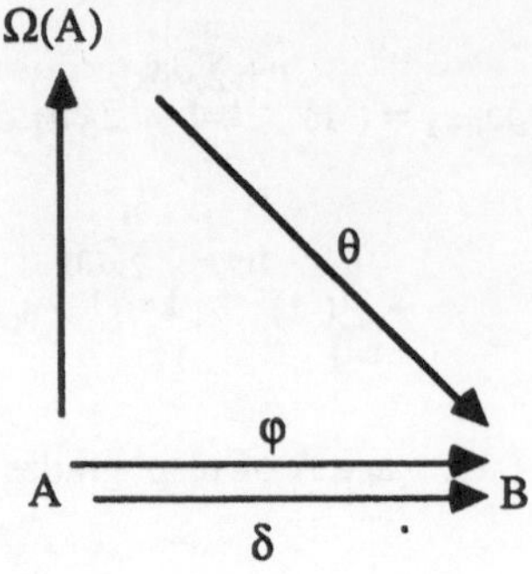

in fact given as follows

(1,11) $$\begin{cases} \theta\{(\lambda\tilde{1}+a_0)da_1 \dots da_n\} = [\lambda+\varphi(a_0)]\ \delta(a_1) \dots \delta(a_n) \\ a_0,\ a_1,\ \dots, a_n \in A \end{cases}$$

[1.4] In addition to being a Z/2-graded complex algebra Ω(A) *possesses a differential d then obtained as follows* :

(1,12) $$d\{(\lambda\tilde{1} + a_0)da_1 \dots da_n\} = da_0\ da_1 \dots da_n;\ a_0, a_1 \dots a_n \in A,$$

this definition implying

(1,13) $$\begin{cases} d(\omega_1,\omega_2) = (d\omega_1)\omega_2 + (-1)^{\partial\omega_1}\omega_1 d\omega_2 \\ \omega_1,\omega_2 \in \Omega(A) \quad , \quad \omega_1 \text{ of grade } \partial\omega_1 \end{cases}$$

and

(1,14) $$d^2 = 0$$

The symbol (1,4) now represents the product of a factor $a_0 \in \tilde{A}$ times n factors da_k obtained by applying the differential d to $a_k \in A$. Note that we then have

(1,15) $$\Omega(A) = A\Omega(A) \oplus d\Omega(A)\,,$$

(with the first (resp. second) direct summand generated by elements of the first (resp. second) line (1,3)[6])

[1.5] We now sketch our "calculus within the differential envelope", based upon the following linear operators of $\Omega(A)$. *For* $\omega \in \Omega(A)$ *and* $a \in A \subset \Omega(A)$ *of respective grades* $\partial\omega$ *and* ∂a *we set* :

(1,16) $$\beta(\omega da) = (-1)^{\partial\omega}\,[\omega,a] \quad , \quad \beta a = 0 \quad ,$$

where [,] *denotes a graded commutator* . This operator β will serve us to define complactly the Hochschild and cyclic boundary operators (cf.(2,3) and (A,1) below). β is *in fact the difference*

(1,17) $$\beta = \beta' - \alpha \quad ,$$

of the acyclic [7] β' :

(1,18) $$\beta'(\omega da) = (-1)^{\partial\omega}\,\omega a \quad , \quad \beta' a = 0 \quad ,$$

and the **"flip-over term"**[8]

(1,19) $$\alpha(\omega da) = (-1)^{(1+\partial a)\partial\omega}\, a\omega \quad , \quad \alpha a = 0$$

related as follows to the **"cyclic permuter"** λ :

(1,20) $$\lambda = \alpha d \quad ,$$

---

[6] One shows that $d\Omega(A)$ is both the kernel and the image of d (in other terms $d\omega=0$, $\omega \in \Omega(A)$ entails $\omega = d\psi \in \Omega(A)$).

[7] In the sense that $\beta'\Omega(A)$ is both the kernel and the image of β' (in other terms $d\beta' = 0, \omega \in \Omega(A)$ entails $\omega = \beta'\psi \in \Omega(A)$.

[8] The names "flip-over term" and "cyclic permuter" refer to the action of the transpose on multilinear forms, cf.(A,5) and (E,28).

whose direct definition reads

(1,21) $$\lambda(\omega da) = (-1)^{(1+\partial a)(1+\partial \omega)}\, a d\omega \;, \quad \lambda a = a$$

Defining further [9]

(1,22) $$\begin{cases} \gamma(\omega da) = [\omega, da] \\ \rho = 1-\gamma \end{cases} \;, \quad \gamma a = 0 \;,$$

we have the following identities [10]

(1,23) $$\beta^2 = \beta'^2 = 0$$

(1,24) $$\beta' d + d\beta' = 1$$

(1,25) $$\beta d + d\beta = \gamma$$

(1,26) $$\lambda + d\alpha = \rho$$

the second of which shows the above-mentioned acyclicity of both β' and d ; whilst (1,26) encompasses the fact (see Appendix A) that cyclic cocycles of A coincide with closed graded traces of Ω(A). *Defining also* a **cyclic symmetrizer** A *by*

(1,27) $$A = \sum_{k=0}^{n} \lambda^n \text{ on } \Omega(A)^n \,,$$

the complementary projections ε *and* 1−ε *on the subspaces* AΩ(A) , resp. dΩ(A) of Ω(A) (cf. 1,15) *are now given by*

---

[9] The reader mainly interested in results will only need the italics portion of the remainder of this section.

[10] For the proof of these identities and the identities below we refer to Chapter 2 of [7].

(1,28) $$\begin{cases} \varepsilon = \alpha d \rho^{-1} = \lambda \rho^{-1} = \lambda^{N+1} = \lambda \rho^{N} \\ 1-\varepsilon = d\alpha\rho^{-1} = A(1-\lambda) = (1-\lambda)A \end{cases}$$

Apart from the operators d,β and α (with subordinates λ, A and γ) *the most important operator of the theory is*

(1,29) $$\mathbb{B} = \mathbb{B}_0 A ,$$

where

(1,30) $$\mathbb{B}_0 = (1-\lambda)\sigma ,$$

with

(1,31) $$\sigma(\omega) = (-1)^{\partial\omega} \omega d1 ,$$

in other terms

(1,30a) $$\mathbb{B}_0\omega = 1\, d\omega + (-1)^{\partial\omega} \omega d1.$$

The operator $\mathbb{B}$ *has a vanishing square*

(1,32) $$\mathbb{B}^2 = 0$$

and anticommutes with $\beta\varepsilon$:

(1,33) $$\mathbb{B}\beta\varepsilon + \beta\varepsilon\mathbb{B} = 0$$

The above operators fulfill a number of identities for which we refer to Chapter 2 of [7] (cf. [7.2] - the reader will also find a partial list in Appendix A of [6]). We conclude this section by noting that *setting*

(1,34) $$\Lambda = \frac{A}{N+1}\ ,\quad \Lambda^2 = 1 - \Lambda$$

where $N|_{\Omega(A)^n} = n$, *the maps* $\phi \to \phi o \Lambda$ *and* $\phi \to \phi o \Lambda^{\perp}$ *are complementary projections of* C *onto* C^*_λ , *resp. the set of* $\phi \in C$ *such that* $\phi o A = 0$, *the latter being of the type* $\phi_1(1-\lambda)$ *with* $\phi_1 \in C$ *given by* $\varphi_1 = u(\lambda)\varphi$, *where*

(1,35) $$u(\lambda) = \begin{cases} (n+1)^{-1} \sum_{k=0}^{n-1} (n-k)\lambda^k & \text{on } \Omega(A)^n \\ 1 & \text{on } \Omega(A)^\circ \end{cases}$$

The following identities are listed for future reference

(1,36) $$\varepsilon\beta\varepsilon = \beta\varepsilon$$

(1,37) $$(\beta\varepsilon)^2 = 0$$

(1,38) $$\alpha\rho^{-1}\varepsilon = 0$$

(1,39) $$\alpha\sigma\varepsilon = \varepsilon$$

(1,40) $$1-\lambda = \beta\, \mathbb{B}_0 + \mathbb{B}_0\beta'$$

(1,41) $$\beta(1-\lambda) = (1-\lambda)\beta'$$

(1,42) $$\begin{cases} \mathbb{B}_0\varepsilon = \varepsilon\mathbb{B}_0 = \varepsilon\mathbb{B}_0\varepsilon \\ \mathbb{B}\varepsilon = \varepsilon\mathbb{B} = \varepsilon\mathbb{B}\varepsilon \end{cases}$$

(1,43) $$\beta\,(\mathbb{B}_0 - d) + \mathbb{B}_0 - d)\beta'$$

§ 2 THE ENTIRE CYCLIC COHOMOLOGY OF A $\mathbb{Z}/2$-GRADED BANACH ALGEBRA

From now on we take $A = A^0 \oplus A^1$ to be a unital $\mathbb{Z}/2$-graded Banach algebra with unit 1 and Banach norm $\| \ \|$. We denote by $\Omega(A)^*$ the algebraic dual (vector space of linear forms) of $\Omega(A)$, with $\Omega(A)^{*n}$ the subset of elements of $\Omega(A)^*$ vanishing outside $\Omega(A)^n$.

[2.1] A **closed linear form** of $\Omega(A)$ is a $\varphi \in \Omega(A)^*$ such that $\varphi \circ d = 0$. We denoted by C the set of closed linear forms of A[11]

(2,1) $$C = \bigoplus_{n \in N} C^n$$

where

(2.2) $$C^n = C \cap \Omega(A)^{*n}$$

The linear operators b and B of C are defined as the following transpose :

(2,3) $$b\varphi = \varphi o \beta \varepsilon$$

$$, \varphi \in \Omega(A)^* ,$$

(2,4) $$B\varphi = \varphi o \mathbb{B} \varepsilon$$

where $\beta, \mathbb{B}$ and ε are the operators (1,16), (1,29) and (1,28). b and B are respectively of grade -1 and +1, thus map C^{ev} (C^{odd}) into C^{odd} (C^{ev}), where

(2,5) $$\begin{cases} C^{ev} = \bigoplus_{n \in N} C^{2n} \\ C^{odd} = \bigoplus_{n \in N} C^{2n+1} \end{cases}$$

These operators fulfill [12]

(2,6) $$b^2 = B^2 = bB + Bb = 0$$

Consequently, if we define

(2,7) $$\Delta = d_1 + d_2$$

---

[11] We shall write C(A) for C whenever the algebra A needs to be specified.

[12] As follows from (1,23), $\beta\varepsilon = \varepsilon\beta\varepsilon$, (1,32) and (1,33).

with d_1 and d_2 the following regauged forms of b and B[13] :

(2,8) $$\begin{cases} d_1\phi = (n+1)b\phi \\ d_2\phi = \frac{1}{n}B\phi \end{cases} ,\phi \in C^n,\ n \in \mathbb{N},$$

giving rise to the analogous relation

(2,6a) $$d_1^2 = d_2^2 = d_1d_2 + d_2d_1 = 0,$$

we get a linear operator Δ of C, acting from $C^{\text{even}}_{\text{odd}}$ to $C^{\text{odd}}_{\text{even}}$, and of vanishing square :

(2,9) $$\Delta^2 = 0 \ .$$

Δ produces a periodic complex

(2,10) $$\to C_\varepsilon^{ev} \overset{\Delta}{\to} C_\varepsilon^{odd} \overset{\Delta}{\to} C_\varepsilon^{ev} \to$$

where the **even** (resp. **odd**) **entire cochains** are

(2,10a) $$C_\varepsilon^{ev} = \left\{ (\phi_{2n}) \subset C^{ev} \ ; \ \sum_{n=0}^{\infty} \|\phi_{2n}\| \frac{z^n}{n!} \text{ is entire} \right\} ,$$

resp.

(2,11a) $$C_\varepsilon^{ev} = \left\{ (\phi_{2n+1}) \in C^{odd} \ ; \ \sum_{n=0}^{\infty} \|\phi_{2n+1}\| \frac{z^n}{n!} \text{ is entire} \right\}$$

with

(2,11b) $$C_\varepsilon = C_\varepsilon^{ev} \oplus C_\varepsilon^{odd} \ .$$

Denoting by $\mathcal{Z}$, resp. $\mathcal{B}$ the following sets of algebraic cocycles, resp. coboundaries :

---

13 This definition makes sense on C° on which B vanishes.

$$(2.12) \qquad \begin{cases} \mathcal{Z} = \mathrm{Ker}\ \Delta \\ \mathcal{B} = \mathrm{Im}\ \Delta \end{cases}$$

the sets of **even** (resp. **odd**) **entire cocycles** are

$$(2.13) \qquad \begin{cases} Z^{ev}_{\varepsilon} = \mathcal{Z}^{ev}_{\varepsilon} \cap C^{ev}_{\varepsilon} \ (\text{resp. } Z^{odd} = \mathcal{Z} \cap C^{ev}_{\varepsilon}) \\ Z^{\varepsilon}_{\varepsilon} = Z^{ev}_{\varepsilon} \oplus Z^{odd}_{\varepsilon} \end{cases}$$

the sets of **even** (resp. **odd**) **entire boundaries** being

$$(2.14) \qquad \begin{cases} B^{ev}_{\varepsilon} = \Delta(C^{ev}_{\varepsilon}) \ (\text{resp. } B^{odd}_{\varepsilon} = \Delta C(^{odd}_{\varepsilon})) \\ B^{\varepsilon}_{ev} = B^{ev}_{\varepsilon} \oplus B^{odd}_{\varepsilon} \end{cases}$$

with $H^{ev}_{\varepsilon} = Z^{ev}_{\varepsilon} / B^{ev}_{\varepsilon}$, resp. $H^{odd}_{\varepsilon} = Z^{odd}_{\varepsilon} / B^{odd}_{\varepsilon}$, the corresponding **even** (resp. **odd**) **entire cyclic cohomology groups.**

[2.2] Remark : Note that *the even and odd parts of any (entire) cyclic cocycle, resp. boundary, is itself an (entire) cyclic cocycle, resp. boundary, this holding for all three Z/2-gradings at hand : the total grading, the intrinsic grading, and the* ℕ*-parity . One has the decomposition*

$$(2,15) \qquad \begin{cases} Z^{\substack{ev \\ odd}}_{\varepsilon} = Z^{\substack{ev,+ \\ odd}}_{\varepsilon} \oplus Z^{\substack{ev,- \\ odd}}_{\varepsilon} \\ B^{\substack{ev \\ odd}}_{\varepsilon} = B^{\substack{ev,+ \\ odd}}_{\varepsilon} \oplus B^{\substack{ev,- \\ odd}}_{\varepsilon} \\ H^{\substack{ev \\ odd}}_{\varepsilon} = H^{\substack{ev,+ \\ odd}}_{\varepsilon} \oplus H^{\substack{even- \\ odd}}_{\varepsilon} \end{cases}$$

where ± denotes an intersection with $\Omega(A)^{*\pm}$. *Analogous properties hold for the sets* $\mathcal{Z}_{\varepsilon}^{\mathrm{ev}/\mathrm{odd}}$ *and* $\mathcal{B}_{\varepsilon}^{\mathrm{ev}/\mathrm{odd}}$ *of algebraic cocycles, resp. boundaries.*

These properties are immediate consequences of the relations $d\theta = -\theta d$, $d\theta_0 = \theta_0 d$, $\Delta\theta = \theta\Delta$, $\Delta\theta_0 = \theta_0\Delta$, where θ, resp. θ_0 denote the grading involutions of $\Omega(A)$ yielding its total, resp. intrinsic grading.

§ 3 THE NORMALIZED ENTIRE COCYCLES AS REPRESENTING ENTIRE CYCLIC COHOMOLOGY

We now exhibit a subclass of entire cyclic cocycles, the so called normalized cyclic cocycles, which represent entire cyclic cohomology (i.e. whose cohomology classes encompass the entire cohomology groups) and which enjoy a priviledged "geometrical interpretation", as discussed in the following section and in section 6 below. A $\phi = (\phi_n) \in \mathcal{Z}$ is called **normalized** whenever $B_0\phi$ is already cyclic, i.e. whenever $B_0\phi = \lambda B_0\phi$, or[14]

(3,1) $$\phi \mathbb{B}_0 \Lambda^{\perp} = 0$$

in other words

(3,1a) $$B_0 \phi_m = \frac{1}{m} B\phi_m \ , \ m \geq 1 .$$

We denote by $\mathcal{Z}_V$, resp. $\mathcal{Z}_V^{ev}$, $\mathcal{Z}_V^{odd}$ the sets of normalized elements of $\mathcal{Z}$, resp. $\mathcal{Z}_V^{ev}$, $\mathcal{Z}_V^{odd}$.

[3.1] Proposition. *For every* $\psi \in \mathcal{Z}$ there is a $\psi' \in \mathcal{Z}_V$ such that $\psi' - \psi \in \mathcal{B}$, namely [15]

(3,2) $$\psi' = \psi - \varphi\beta\varepsilon \quad \text{with} \quad \varphi = \psi \, \mathbb{B}_0 u(\lambda)$$

Moreover, if $\psi \in \mathcal{Z}_\varepsilon$, *one has* $\phi \in C_\varepsilon$, *thus,* $\psi' - \psi \in B_\varepsilon$: *hence each entire cyclic cohomology class can be represented by a normalized entire cocycle.*

As one would expect, the proof of this Proposition rests upon (a slight extension of) one of the fundamental technical lemmas of the original versions of cyclic cohomology (Lemma [6.3] of [1], adapted to the $\mathbb{Z}/2$-graded frame as Lemma 36 of [3] - cf [A.8] and [A.9] below)[16].

---

[14] cf.(1,34) for the definition of $\Lambda^{\perp}$

[15] thus one has $\psi B_0 \Lambda^{\perp} = \varphi(1-\lambda)$

[16] For a detailed proof see Section [1] of [6]

We now discuss a "regauging" $\phi \to \tilde{\phi}$ of the cochains $\phi \in C$ motivated by the normalization condition (3,1a) and useful in formulating Prop. [4.1].

[3.2] Remark : *Let the bijection* $C \ni \phi \leftrightarrow \tilde{\phi} \in C$ *be defined as follows*

$$(3,3) \qquad \begin{cases} \phi_{2n} = (1)^n (2n-1) \ldots\ldots\ldots 3.1\, \tilde{\phi}_{2n} \,, & \phi \in C^{ev} \\ \phi_{2n+1} = (1)^n (2n) \ldots\ldots\ldots 4{\cdot}2\, \tilde{\phi}_{2n+1} \,, & \phi \in C^{odd} \,, \\ (\phi_0 = \tilde{\phi}_0 \,,\ \phi_1 = \tilde{\phi}_1\,) & \end{cases}$$

$\phi \in Z$ *is normalized iff one has*

$$(3,4) \qquad b\,\tilde{\phi}_m = B_0 \tilde{\phi}_{m+2} \quad , m \in \mathbb{N}$$

i.e.

$$(3,4a) \qquad \tilde{\phi}\beta\varepsilon = \tilde{\phi}\, \mathbb{B}_0$$

the definition of the entire cochains reading

$$(3,5) \qquad \begin{cases} C^{ev}_{\varepsilon} = \left\{ (\tilde{\phi}_{2n}) \ ; \ \sum_{n=0}^{\infty} \|\tilde{\phi}_{2n}\| (2z)^n \quad \text{is entire} \right\} \\ C^{odd}_{\varepsilon} = \left\{ (\tilde{\phi}_{2n+1}) \ ; \ \sum_{n=0}^{\infty} \|\tilde{\phi}_{2n+1}\| (2z)^n \quad \text{is entire} \right\} \end{cases}$$

(Note that $\phi \in C$ *belongs to* Z_ν *iff one has (3,4a) and* $\phi \mathbb{B}_0 \lambda = \phi \mathbb{B}_0$ $\phi \in C$*)*

Analogously to the fact noted for general (entire) cocycles in Remark [2.2] we have

[3.3] Remark. *The set of normalized (entire) cocycles is a graded subspace w.r.t.* N-*parity, total grading, and intrinsic grading*. (Immediate consequence of the relations $\mathbb{B}_0\theta = -\theta\,\mathbb{B}_0$, $\lambda\theta = \theta\lambda$; and $\mathbb{B}_0\,\theta_0 = \theta_0\,\mathbb{B}_0$, $\lambda\,\theta_0 = \theta_0\,\lambda$).

§ 4 THE NORMALIZED CYCLIC COCYCLES OF A AS 1-PARATRACES OF $\Omega(A)$

The interest of the normalized cyclic cocycles resides in their interpretation as 1-paratraces of the differential envelope as we now show (and further as traces of the Cuntz or Zekri algebras - cf. Section 7 below).

We begin with a purely algebraic statement :

[4.1] Proposition - *Assigning to the* $\psi = (\psi_n) \in C$ *the following linear forms* μ_ψ *on* $\Omega(A)$:

$$(4.1) \qquad \begin{cases} \mu_\psi(a_0 da_1 \dots da_n) = \psi_n(a_0 da_1 \dots da_n) \\ \mu_\psi(da_1 \dots da_n) = \psi_n(\mathbb{B}_0(a_1 da_2 \dots da_n)) \end{cases} \qquad , a_0, a_1, \dots, a_n \in A$$

we have that [17]

(i) *for each* $\phi \in Z_\nu$, *with* $\tilde{\phi}$ *as in* [3.2] , $\mu_{\tilde{\phi}}$ *is a 1-paratrace of* $\Omega(A)$

(ii) *conversely each 1-paratrace* μ *vanishing on* dA *of* $\Omega(A)$ *determines* a $\phi \in Z_\nu$ s.t.

$$(4.2) \qquad \tilde{\phi}(\omega) = \mu(\omega) \quad , \omega \in Ad\Omega(A).$$

This yields a (ℕ*-parity conserving*) *bijection between the normalized element of* Z *and the 1-paratraces of* $\Omega(A)$ *vanishing on* dA.

Sketch of the proof [18] : in order to prove (i) it suffices (cf.(C.3) below) to check the vanishing of $\mu_{\tilde{\phi}}$ on the 1-parabrackets

$$(4.3) \begin{cases} \{\omega, x\}_1 = (-1)^{\partial\omega}(\beta - d)(\omega dx) , \\ \{\omega, dx\}_1 = \gamma(\omega dx) = -(\beta - d)^2(\omega dx) \end{cases} \qquad \begin{cases} \omega \in \Omega(A) \text{ of grade } \partial\omega \\ x \in A \end{cases}$$

---

[17] For the notion of 1-paratrace see Appendix C

[18] For a detailed proof, see section [2] of [6].

This is done by checking the identities $\mu_{\tilde{\phi}} \circ (\beta\text{-}d) = 0$ and $\mu_{\tilde{\phi}} \circ \gamma = 0$, using the calculus within $\Omega(A)$.

(ii) $\tilde{\phi}$ in (4.2) is shown to be cyclic ($\Leftrightarrow 0 = \mu \varepsilon B_0 d = \mu \varepsilon B_0 \gamma$ cf. Appendix A, Lemma [A.1]) and to fulfill (3.4a), using again the calculus within $\Omega(A)$. The last statement in [4.1] procedes from the fact that the 1-paratraces μ of $\Omega(A)$ are determined by their restrictions to $A\Omega(A)$, owing to the equality $\mu\,(d\omega) = \mu(B_0\omega)$, $\omega \in A\Omega(A)$.

For applying the algebraic result [4.1] to the normalized cyclic cocycles it is convenient to regauge the $\mu_{\tilde{\phi}}$, noticing that the map $\mu \leftrightarrow \nu$ given by

$$(4,4) \qquad \begin{cases} \nu_{2n} = 2^n \mu_{2n} \\ \\ \nu_{2n+1} = 2^n \mu_{2n+1} \end{cases} \quad , n \in \mathbb{N}$$

maps bijectively the 1-paratraces μ onto the $\frac{1}{2}$- paratraces ν of $\Omega(A)$. We then have

[4.2] Theorem *There is a bijection between the even, resp. odd.* $\phi \in Z_\nu$ *and the even, resp. odd.* $\frac{1}{2}$ *- paratraces* ν_ϕ *of* $\Omega(A)$ *, given by*

$$(4,5) \qquad \begin{cases} (\nu_\phi)_{2n} = 2^n (\mu_{\tilde{\phi}\,2n}) \\ \\ \text{resp.} (\nu_\phi)_{2n+1} = 2^n (\mu_{\tilde{\phi}})_{2n+1} \end{cases} \quad , n \in \mathbb{N}$$

This bijection maps $Z_\nu^{ev} \cap Z_\nu$, resp. $Z_\nu^{odd} \cap Z_\nu$, *onto those* $\frac{1}{2}$ *paratraces* ν *of* $\Omega(A)$ *for which the power series*

$$(4,6) \qquad \begin{cases} \sum_{n=1}^{\infty} \|(\nu_\phi)_{2n}\| z^n \\ \\ \text{resp.} \sum_{n=1}^{\infty} \|(\nu_\phi)_{2n}\| z^n \end{cases} \quad , n \in \mathbb{N} ,$$

represent entire functions - in other terms onto the continuous paratraces ν_ϕ of $\Omega(A)$ *topologized by the following norms* $\|\ \|_\pi$, $r\in R$:

(4,7) $$\|\sum_{k=0}^{\infty}\omega_k\|_r = \sum_{k=0}^{\infty} r^k \|\omega_k\|_\pi$$

(π the projective tensor norm)

We note that the relations (4,1), (4,4) imply that

(4,8) $$\begin{cases}\phi_{2n} = (-1)^n \dfrac{\Gamma(n+\frac{1}{2})}{\sqrt{\pi}}\{\nu_\phi|_{A\Omega(A)}\}_n \\ \phi_{2n+1} = (-1)^n\, n!\, \{\nu_\phi|_{A\Omega(A)}\}_{2n+1}\end{cases}, n\in \mathbb{N}$$

where

(4,9) $$\frac{\Gamma(n+\frac{1}{2})}{\sqrt{\pi}} = \begin{cases}(n-\frac{1}{2})\cdot\cdot\cdot\cdot\cdot\frac{3}{2}\ \frac{1}{2} & , n \geq 1 \\ 1 & , n = 0\end{cases}$$

[4.3] Remark. *The bijection between the normalized cocycles ϕ of* A *and the $\frac{1}{2}$-paratraces ν_ϕ of $\Omega(A)$ in Theorem [4.2] holds separately for the even and odd parts of ϕ and ν_ϕ with respect to the N -, and also the total and the intrinsic grading* (as follows from Remarks [3.3] and [6.3]).

[4.4] Remark

(i) *All 1-parabrackets in* $\Omega(A)$ *are linear combinations of parabrackets of the types* (4.3).

(ii) *Consequently* Coker $(\beta\text{-d})$ *consist of 1-paratraces of* $\Omega(A)$ *(in fact of those paratraces vanishing on dA). One has equality of the following subsets of* $\Omega(A)^*$

(4,10) $$\{\mu_{\tilde\phi}, \phi\in Z_\nu\} = \mathrm{Coker}(\beta\text{-d}) = \{1\text{-paratraces of } \Omega(A)\} \cap (dA)^\perp$$

(iii) *For* μ *a one-paratrace of* $\Omega(A)$, *the canonical projection* $\mu^\perp$ *of* μ *on* $(dA)^\perp$ *is again a 1-paratrace of* $\Omega(A)$.

Proof : (i) follows from (C.3) in Appendix C, and implies the second equality (4,10). Proof of the first equality (4.10) : The inclusion $\subseteq$ follows from (4,1) and (A,43) ; opposite inclusion : noticing that for $\omega \in \Omega(A)$ of total grade $\partial\omega$:

$$\mathbb{B}_0\omega = (-1)^{\partial\omega} \{\omega, d1\}_1 + d\,(1\omega), \tag{4,11}$$

it follows that *each 1-paratrace* μ *of* $\Omega(A)$ *fulfills* $\mu d = \mu\varepsilon\,\mathbb{B}_0$ *with* $\mu\varepsilon\mathbb{B}_0$ *cyclic* (since, applying [A.1] (ii), $\mu\,\varepsilon\,\mathbb{B}_0 d = \mu\,\mathbb{B}_0\,\varepsilon\,d = 0$ and $\mu\varepsilon\mathbb{B}_0\gamma = \mu d\gamma = \mu\gamma d = 0$ by (4.3) and $\gamma dx = 0$, $x \in A$). Moreover, for $\mu \in$ Coker $(\beta\text{-}d)$, $\mu\varepsilon(\beta\varepsilon\text{-}\mathbb{B}_0) = \mu(\beta\text{-}d)\varepsilon = 0$. Finally (iii) follows from (4.3).

[4.5] Remark

For $\phi \in Z_\nu$ *we have the equivalences*

$\tilde{\phi}\,\mathbb{B}_o = 0 \Leftrightarrow \tilde{\phi}\,\mathrm{o}\mathbb{B} = 0 \Leftrightarrow \tilde{\phi}\,\mathrm{o}\,(1\text{-}\lambda) = 0 \Leftrightarrow \mu_{\tilde{\phi}}\,\mathrm{o}\,(1\text{-}\lambda) = 0 \Leftrightarrow \mu_{\tilde{\phi}}\,\mathrm{o}d = 0.$

§ 5. THE CUNTZ ENVELOPE QA AND ZEKRI ALGEBRA $\mathcal{E}A$ OF A $\mathbb{Z}/2$-GRADED COMPLEX ALGEBRA A

Let A be a $\mathbb{Z}/2$-graded complex algebra, and let $\Omega(A)$ be the differential envelope of A with differential d (cf. (1,2)), and grading involution θ :

$$\theta\omega = (-1)^{\partial\omega} \quad , \quad \omega \in \Omega(A)^+ \cup \Omega(A)^- , \tag{5.1}$$

where ∂ is the total grading (1,8). The **Cuntz envelope** Q_tA, and **Zekri algebra** $\mathcal{E}_tA, t\in\mathbb{C}$, are obtained by applying the general construction of Appendix B to the $\mathbb{Z}/2$-graded differential algebra $(\Omega(A),\theta,d)$

$$\begin{cases} Q_tA \doteq Q_t\Omega(A) & ,QA \doteq Q_1\Omega(A) \\ \mathcal{E}_tA \doteq \mathcal{E}_t\Omega(A) & ,\mathcal{E}A \doteq \mathcal{E}_1\Omega(A) \end{cases} \tag{5,2}$$

Q_tA *hence coincides with* $\Omega(A)$ *as a vector space, but has a different structure of a* $\mathbb{Z}/2$-*graded algebra, with unchanged differential d, but modified grading involution :*

$$\tau\omega = (-1)^{\partial\omega}\,(\omega - d\omega) \quad , \quad \omega \in \Omega(A)^+ \cup \Omega(A)^- \tag{5,3}$$

and modified product :

$$\omega \underset{t}{*} \omega' = \begin{cases} \omega\omega' & ,\omega \in \Omega(A)^+ \\ \omega\omega' + \omega d\omega' & ,\omega \in \Omega(A)^- \end{cases} \tag{5,4}$$

with the property :

$$d(\omega \underset{t}{*} \omega') = (d\omega) * \omega' + (-1)^{\partial\omega}\,\omega * d\omega' - (-1)^{\partial\omega}\,(d\omega) * d\omega' , \quad \begin{cases} \omega \in \Omega(A)^+ \cup \Omega(A)^- \\ \omega' \in \Omega(A) \end{cases} \tag{5,5}$$

i.e.

$$d(\omega \underset{t}{*} \omega') = (d\omega) * \omega' + (\tau\omega) * d\omega' \ , \quad \omega, \omega' \in QA \ . \tag{5,5a}$$

$\mathcal{E}_t A = Q_t A \underset{\tau}{\times} (\mathbb{Z}/2)$ can be drescribed as the set of formal sums

$$\omega_0 \oplus \omega_1 f \qquad ,\omega_0, \omega_1 \in Q_t(A) \tag{5,6}$$

with (associative) multiplication specified by the rules

$$\begin{cases} f^2 = 1 \\ f\ \omega\ f = \tau\,\omega \end{cases} . \tag{5,7}$$

This way of looking at $\mathcal{E}_t A$ consists in embedding it within[19]

$$\tilde{\mathcal{E}}_t A = Q_t \tilde{\Omega}(A) \underset{\tau}{\times} (\mathbb{Z}/2) = \mathcal{E}_t A \oplus \mathbb{C}\, 1_{\tilde{\mathcal{E}}A} \oplus \mathbb{C} f \tag{5,8}$$

where

$$\tilde{1}_{\tilde{\mathcal{E}}A} = (\tilde{1}, 0) \ , \ f = (0, \tilde{1}) \tag{5,9}$$

As it is the case for $\Omega(A)$, $Q_t A$ is characterized by a universality property, namely[20][21] .

[5.1] Proposition *Let $Q_t A = Q_t \Omega(A)$, $t \in \mathbb{C}$, $A = A^0 \oplus A^1$ a $\mathbb{Z}/2$-graded complex algebra ; and let B be a complex algebra.*
There is a bijection between
(i) *the pairs (α, δ) of linear maps $: A \to B$ fulfilling*[22]

---

[19] We recall that $\tilde{\Omega}(A) = \mathbb{C}\tilde{1} \oplus \Omega(A)$

[20] For $t = 0$ one gets the universality property of $\Omega(A)$ described in section 1.

[21] The hurried reader may skip the remainder of this section

[22] where $Q_t A$ is equipped with the product $\underset{t}{*}$

$$(5{,}10)\qquad \alpha(aa') = \begin{cases} (\alpha a)\ (\alpha a') & ,a \in A^0 \\ (\alpha a)(\alpha a') - t(\alpha a)(\delta a') & ,a \in A^1 \end{cases}$$

and

$$(5.11)\qquad \delta(aa') = \begin{cases} (\delta a)(\alpha a') + (\alpha a)(\delta a') - t(\delta a)(\delta a') & ,a \in A^0 \\ (\delta a)(\alpha a') - (\alpha a)(\delta a') & ,a \in A^1 \end{cases}$$

(ii) *the homomorphisms* $\nu :\ Q_tA \to B$ *of complex algebras* [21]

(5,12)

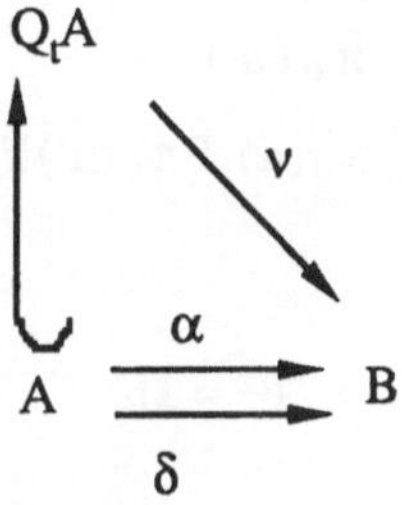

This bijection is specified by the relations :

$$(5{,}13)\qquad \begin{cases} \alpha a = \nu a \\ \delta a = \nu da \end{cases} \qquad ,a \in A$$

implying

$$(5{,}14)\qquad \begin{cases} \nu(a_0 da_1 \ldots\ldots da_n) = (\alpha a_0)(\delta a_1) \ldots (\delta a_n) \\ \nu(da_1 \ldots\ldots da_n) = (\delta a_1) \ldots (\delta a_n) \end{cases} \qquad ,a_0, a_1, \ldots a_n \in A$$

Note that, in the particular case of trivially graded A ($A=A^0, A^1=\{0\}$) and for t = 1, we obtain a bijection between the homomorphism : ν $QA \to B$ and the pairs $(\alpha,\alpha') \in \mathrm{Hom}(A,B)$ where $\alpha'=\alpha-\delta$.

The next result is a universality property of $\mathcal{E}_t A$, $t \neq 0$, useful for constructing the characters of θ-summable Fredholm modules.

[5.2] **Proposition** : *Let* $\tilde{\mathcal{E}}_t A = Q_t \tilde{\Omega}(A) \underset{\tau}{\times} Z/2$, $A = A^0 \oplus A^1$ *a Z/2-graded complex algebra ; and let* B *be a unital complex algebra with unit* 1_B. *There is a bijection between :*

(i) *the unital homomorphisms* $\pi : \tilde{\mathcal{E}}_t A \to B$ *of complex algebras*

(ii) *the pairs* (π_0, F) *of a linear map* $\pi_0 : A \to B$ *and a* $F \in B$ *such that*

$$\pi_o(aa') = \begin{cases} \pi_0(a)\, \pi_0(a') & , a \in A^0 \\ (-1)^{\partial a'}\, \pi_0(a)\, F \pi_0(a') F & , a \in A^1 \end{cases} \tag{5,15}$$

and

$$F^2 = 1_B \tag{5,14}$$

This bijection is given as follows : given π, *one has* [23]

$$\begin{cases} \pi_0 = \pi \big|_{A \in (Q_t \tilde{\Omega}(A),0)} \\ F = \pi\,(f) \end{cases} \tag{5,16}$$

and, given the pair (π_0, F), *one has, for* $\omega_0, \omega, \in Q_t\tilde{\Omega}(A)$

$$\pi(\omega_o + \omega, f) = \bar{\pi}_o(\omega_o) + \bar{\pi}_o(\omega_1)\, F \tag{5,17}$$

where $\bar{\pi}_0(\tilde{1}) = 1_B$ *and , for* $a_0, a_1, \ldots, a_n \in A$:

[23] where $f = (0,\tilde{1})$, cf. (5,6)

(5,18) $$\begin{cases} \bar{\pi}_0\,(a_0 da_1 \ldots da_n) = \pi_0(a_0)\,\Delta\,(a_1) \ldots \Delta(a_n) \\ \bar{\pi}_0(da_1 \ldots da_n) = \Delta(a_1) \ldots \Delta(a_n) \end{cases}$$

with

(5,19) $$\Delta(a) = \frac{1}{t}\left\{\pi_0(a) - (-1)^{\partial a}\, F\pi_0(a)F\right\}$$

Note that, for A trivially graded, we get a bijection between the homomorphisms : $\tilde{\mathcal{E}}_t\ A \to B$, and the pairs of an involution F of B, and a homomorphism $\pi_0 : A \to B$. In this case, and if t = 1, we have

(5,20) $$\Delta(a) = F\,[F,\pi_0(a)]$$

The proof of [5.1) is routine, that of [5.2] proceeds easily from Remark [B.3] (iii) of Appendix B. (making there $\Omega = \tilde{\Omega}(A)$: (5,15) is then the expression of the requirement (B,14)).

[5.3] Remark *Insertion of* (5,19) *in* (5,18) *yields, for* $a_0 \in \tilde{A}$, $a_1, \ldots a_n \in A$:

(5,18a) $$\bar{\pi}(a_0 da_1 \ldots da_n) = t^{-n}\,(-1)^{[n] + \sum_{k\,odd} \partial a_k}\, \pi_0(a_0)\,[F,\pi(a_1)] \ldots [F,\pi(a_n)]$$

where $[2n] = n$, $[2n+1] = n+1$, $n \in \mathbb{N}$ and $\pi_0(\tilde{1}) = 1_B$ *and*

(5,21) $$[F,\pi_0(a)] = F\pi_0(a) - (-1)^{\partial a}\,\pi_0(a)F \ , \quad a \in A^0 \cup A^1 \ ;$$

in particular, for A *trivially graded*, t = 1, *and* n *pair*

(5,22) $$\begin{cases} \bar{\pi}(a_0 da_1 \ldots da_{2n}) = (-1)^n\,\pi_0(a_0)\,[F,\pi(a_0)] \ldots [F,\pi(a_{2n})] \\ \bar{\pi}(a_0) = \pi_0(a_0) \end{cases}$$

where the [,] *now denote commutators* .

§ 6 THE EVEN , RESP. ODD, ENTIRE CYCLIC COCYCLES AS ODD TRACES OF $\mathcal{E}A$, RESP. QA

We are now ready to describe the next useful characterization of the normalized cyclic cocycles.

[6.1] Theorem (i). *There is a bijection between the odd*[24] *elements* ϕ *of* Z_V *and the odd*[25] *traces* T *of QA, given by*

(6.1)
$$T(\omega) = \begin{cases} \frac{1}{2}\nu_\phi(d\omega) & , \omega \in \Omega(A)^+ \\ \nu_\phi(\omega) & , \omega \in \Omega(A)^- \end{cases}$$

(ii) *There is a bijection between the even*[24] *elements* ϕ of Z_V *and the odd*[26] *traces* $\mathbb{T}$ *of* $\mathcal{E}$A *fulfilling* $\mathbb{T} \circ \tau = \mathbb{T}$, *given by*

(6.2)
$$\mathbb{T}(\omega_0,\omega_1) = \begin{cases} \nu_\phi(\omega_1) & , \omega_1 \in \Omega(A)^+ \\ \frac{1}{2}\nu_\phi(d\omega_1) & , \omega_1 \in \Omega(A)^- \end{cases}$$

(iii) *These bijections map* $Z_\varepsilon^{\mathrm{odd}} \cap Z_V$ (*resp.* $Z_\varepsilon^{\mathrm{ev}} \cap Z_V$) *on those traces* T (*resp* .$\mathbb{T}$) *which are continuous for* QA *topologized with the norms (4,7) (respectively* $\mathcal{E}$A *endowed with the corresponding direct sum topology)* .

Considering the bijections in [6.1], [3.1], [3.2] we obtain owing to Remarks [4.3] and [C.3] in Appendix C the following bijection :

With the notation

---

24 $\phi \in Z_V$ is $\begin{smallmatrix}\text{odd}\\ \text{even}\end{smallmatrix}$ whenever it vanishes $\Omega(A)^\pm$, where $\Omega(A)^\pm$ denotes the $\begin{smallmatrix}\text{even}\\ \text{odd}\end{smallmatrix}$ part of $\Omega(A)$ for its (total) grading.

25 odd in the sense that $T \circ \tau = -T$

26 odd in the sense $\mathbb{T} \circ \hat{\tau} = -\mathbb{T}$

(6,3) $\left(Z^{ev}_{\varepsilon v}\right)^{0}_{1} = \{$ $(-1)^N$-even normalized entire cocycles of A of $\genfrac{}{}{0pt}{}{\text{even}}{\text{odd}}$ intrinsic grade $\}$

$= Z^{ev\pm}_{\varepsilon v} \cap Z_v$

(6,4) $\left(Z^{odd}_{\varepsilon v}\right)^{0}_{1} = \{$ $(-1)^N$-odd normalized entire cocycles of A of $\genfrac{}{}{0pt}{}{\text{even}}{\text{odd}}$ intrinsic grade $\}$

$= Z^{odd\pm}_{\varepsilon v} \cap Z_v$

(6,5)[27] $\bar{Tr}(QA)^{0}_{1} = \{$ τ-odd traces of QA of $\genfrac{}{}{0pt}{}{\text{even}}{\text{odd}}$ intrinsic grade $\}$

(6,6) $\bar{Tr}_\tau(\mathcal{E}A)^{0}_{1} = \{$ $\hat{\tau}$-odd traces of QA of $\genfrac{}{}{0pt}{}{\text{even}}{\text{odd}}$ intrinsic grade $\}$

and $a_0, a_1, \ldots, a_{2n+1} \in A$, $n \in \mathbb{N}$, *we have the bijections* :

– $\left(Z^{even}_{\varepsilon v}\right)^0 \ni \phi \leftrightarrow \mathbb{T} \in \bar{Tr}_\tau(\mathcal{E}A)^0$ *given by*

(6,7)
$$\begin{cases} \phi(a_0 da_1 \ldots da_{2n}) = \dfrac{(-1)^n}{\sqrt{\pi}} \Gamma(n+\frac{1}{2})\, \mathbb{T}((0, a_0 da_1 \ldots da_{2n})), n \in \mathbb{N} \\ \phi(a_0) = \mathbb{T}((0, a_0)) \end{cases}$$

– $\left(Z^{odd}_{\varepsilon v}\right)^1 \ni \phi \leftrightarrow \mathbb{T} \in \bar{Tr}_\tau(\mathcal{E}A)^1$ *given by*

(6,8) $$\phi(a_0 da_1 \ldots da_{2n+1}) = (-1)^n\, n!\; \mathbb{T}(0, a_0 da_1 \ldots da_{2n}), \quad n \in \mathbb{N}$$

– $\left(Z^{even}_{\varepsilon v}\right)^1 \ni \phi \leftrightarrow T \in \bar{Tr}(QA)^1$ *given by*

---

[27] Note that, for $\mathcal{T}$ a τ-odd trace of QA (resp. for $\mathbb{T}$ a $\hat{\tau}$-odd τ-even trace of $\mathcal{E}A$) the even and odd part of T (resp. of $\mathbb{T}$) for the intrinsic grading is again of the same nature.

$$(6,9) \qquad \begin{cases} \phi(a_0 da_1 \ldots. da_{2n}) = \dfrac{(-1)^n}{\sqrt{\pi}} \Gamma(n+\frac{1}{2})\ T\ (a_0 da_1 \ldots. da_{2n}), \quad n \in \mathbb{N} \\ \phi(a_0) = T(a_0) \end{cases}$$

– $\left(Z_{\varepsilon v}^{odd}\right)^0 \ni \phi \leftrightarrow T \in \overline{\text{Tr}}\,(QA)^0$ *given by*

$$(6,10) \qquad \phi(a_0 da_1 \ldots. da_{2n+1}) = (-1)^n\, n\,!\ \ T\ (a_0 da_1 \ldots da_2), \qquad n \in \mathbb{N}$$

§ 7 THE CHARACTERS OF θ-SUMMABLE FREDHOLM MODULES

In this section we sketch Alain Connes' construction of normalized even entire cyclic cocycles as the characters of θ-summable Fredholm modules. This construction pertaining to the case of trivially graded Banach algebras A ; we will assume $A = A^0$, $A^1 = \{0\}$ throughout this section. As stated in (6,8) even entire normalized cyclic cocycles ϕ of A are obtained from $\mathfrak{k}$ - odd, τ-even traces $\mathbb{T}$ of $\mathcal{E}A$. Now a general procedure for producing a trace $\mathbb{T}$ of $\mathcal{E}A$ consists in constructing a homomorphism π of the algebra $\tilde{\mathcal{E}}A$ into some unital algebra B endowed with a trace $\mathcal{T}$ and taking the restriction $\mathbb{T}$ to $\mathcal{E}A^-$ of the pull back $\mathcal{T}\circ\pi$[28] . The general form of such homomorphisms π has been described in [5.2] : they are given in the present trivially graded case by pairs (π_0,F) of a $\pi_0 \in \mathrm{Hom}(A,B)$ and an $F\in B$ such that $F^2 = 1_B$: combining (5,17), (5,22) with (6,8) then yields

$$(7,1)\quad \begin{cases} \phi(a_0 da_1 \ldots da_{2n}) = \dfrac{\Gamma(n+\frac{1}{2})}{\sqrt{\pi}}\,\mathcal{T}\big(F\pi_0(a_0)[F,\pi_0(a_1)]\ \ldots\ [F,\pi_0(a_{2n})]\big) \\ \phi(a_0)=\mathcal{T}\big(F\pi_0(a_0)\big) \end{cases}$$

where $a_0,a_1, \ldots, a_{2n}\in A$ and the [,] are commutators.
Note that the trace $\mathbb{T} = \mathcal{T}\circ\pi$ is τ-invariant : indeed one has (cf. (B,15) in Appendix B)

$$(7,2)\quad \mathbb{T}(\tau(\omega_0,\omega_1)) = \mathbb{T}((\tau\omega_0,\omega_1)) = \mathcal{T}(\pi(\tau\omega_1)F) = \mathcal{T}(F\pi(\omega_1)) = \mathcal{T}(\pi(\omega_1)F) = \mathbb{T}((\omega_0,\omega_1))$$

We now sketch the way in which Alain Connes obtains a construction of the above type in § 4 of [1] starting from a θ-summable Fredholm module. We recall that an **unbounded Fredholm module over A** (A a trivially graded Banach algebra) is a couple $(\mathcal{H},D)$ of a graded Hilbert space $\mathcal{H}= \mathcal{H}^0 \oplus \mathcal{H}^1$ with a ∗-respresentation a→ (a) of A by bounded operators of grade zero ; and with an (unbounded) self adjoint opeartor D of

---

[28] Note that the restricition of the trace $\tau \circ \pi$ of $\mathcal{E}A$ to the $\mathfrak{k}$-odd part $\mathcal{E}A^-$ of $\mathcal{E}A$ is a $\mathfrak{k}$-odd trace $\mathbb{T}$ of $\mathcal{E}A$.

$\mathcal{H}$ of grade one[29], such that $(1+D^2)^{-1}$ is compact, and the commutators $[D,(a)]$ are bounded for each $a\in A$. $(\mathcal{H},D)$ is called **θ-summable** whenever e^{-tD^2} is trace class for all $t>0$.

With these data we shall now construct a unital algebra $\tilde{\mathcal{L}}$, a trace $\mathcal{T}$ of $\tilde{\mathcal{L}}$, an involution F of $\tilde{\mathcal{L}}$, and a homomorphism $\pi_0 : A \to \tilde{\mathcal{L}}$, the necessary ingredients of the above construction where $B = \tilde{\mathcal{L}}$.

We first consider the following set $\mathcal{L}$ of operator valued distribution : with $\mathcal{B}(\mathcal{H})$ the set of bounded operators on $\mathcal{H}$ and $\mathcal{S}$ the Schwartz space of fast decreasing smooth functions on the reals, $\mathcal{L}$ consists of the distributions

$$T : \mathcal{S} \to \mathcal{B}(\mathcal{H}) \tag{7,3}$$

such that

(i) Supp. $T\in [0, +\infty[$

(ii) for each $T\in\mathcal{L}$ (there is $r\in\,]0,1[$ and a holomorphic function $t = C_r \to \mathcal{B}(\mathcal{H})$, where $C_r = \bigcup_{\rho>0} \rho\, U_r$, $U_r = \{z\in\mathbb{C}\,;\, |z-1| \leq r\}$, such that (with the obvious meaning)

(a) $t(s) = T(s)$, $s\in\,]0, +\infty[$

(b) there is $\alpha > 0$ s.t.

$$\sup_{z\in\frac{1}{p}U_r} \|t(z)\|_p \leq p^{\alpha},\quad p\in[1,+\infty[\tag{7,4}$$

where $\|\ \|_p$ is the Schatten norm $\|A\|_p = \{T_r|A|^p\}^{\frac{1}{p}}$, $|A| = |A^*A|^{\frac{1}{2}}$, $A\in\mathcal{B}(\mathcal{H})$.

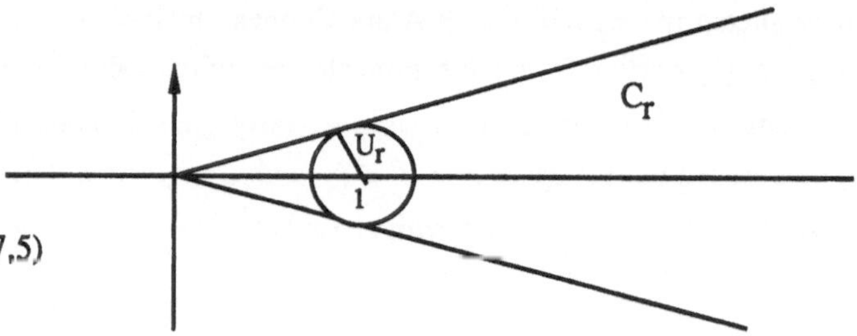

(7,5)

[29] in particular, with ε the grading involution of $\mathcal{H}$ and $\mathcal{D}_D$ the domain of D, we have $\varepsilon\,\mathcal{D}_{\mathcal{D}} = \mathcal{D}_{\mathcal{D}}$.

We denote by $\mathcal{L}^{(\alpha)}$ the subset of $T \in L$ fulfilling (7,4). In particular we have

$$(7,6) \qquad \mathcal{L}^{(0)} = \left\{ T \in \mathcal{L}; \quad \sup_{p \in [1,+\infty[} \ \sup_{z \in \frac{1}{p} U_r} \|T(z)\| < \infty \text{ for some } r > 0 \right\}$$

It easily follows from the above definition that

(i) *for* $T \in \mathcal{L}$, $T(1)$ *is trace-class*

(ii) for each $S \in \mathcal{L}^{(0)}$ we have

$$(7,7) \qquad S(f) = \int_0^\infty S(s)\, f(s) ds \quad , \quad f \in \mathcal{S}$$

Furthermore, with φA, $\varphi \in s'$, $A \in \mathcal{B}(\mathcal{H})$ the the operator-valued distribution

$$(7,8) \qquad (\varphi A)\,(f) = \varphi(f) A \quad , f \in \mathcal{S}$$

we have that

(iii) $\varphi A \in \mathcal{L}$ for A trace-class and φ, carried by $[0, +\infty[$, boundary value of a $\varphi(z)$ holomorphic in some C_r, $r > 0$

(iv) $\varphi A \in \mathcal{L}$ for $A \in \mathcal{B}(\mathcal{H})$ and φ carried by $\{0\}$. In fact each element of $\mathcal{L}$ carried by $\{0\}$ is a sum

$$(7,9) \qquad \sum_{k=1}^{n} \delta^{(k)} A_k \quad , A_k \in \mathcal{B}(\mathcal{H}),\ k = 1, \ldots, n$$

of distributions of this type.

A main consequence (and motivation) of the definition of $\mathcal{L}$ is that $\mathcal{L}$ *is stable under the convolution, the adjoint operation, and the derivation of distributions.* $\mathcal{L}$ is *therefore a complex ∗--algebra.*[30] .The proof of this result is technical (cf. section 3 of [1]) and uses the fact that each $T\in\mathcal{L}$ can be written as the sum of the qth derivative of an $S\in\mathcal{L}^{(0)}$, $q\in\mathbb{N}$, and an element of the form (7,9).

Note that according to (iv) above *we have that*

$$(7,10) \qquad \lambda = \delta'_0 1 \in \mathcal{L}$$

with convolution square root $\lambda^{\frac{1}{2}}(s) = (\pi s)^{-1/2}$ *not included in* $\mathcal{L}$, because failing to fullfil condition (ii) in the above definition of $\mathcal{L}$. In order to define a ∗-algebra with ∗-subalgebra $\mathcal{L}$ and containing $\lambda^{\frac{1}{2}}$, Alain Connes was thus led to *equip the set* $\tilde{\mathcal{L}} = \mathcal{L} \oplus \mathcal{L}$ of formal sums

$$(7,11) \qquad (T,S) = T + S\ \lambda^{\frac{1}{2}} \qquad , T, S\in \mathcal{L}$$

with the product

$$(7,12) \qquad (T_1,S_2)\,(T_2,S_2) = (T_1 * T_2 + \lambda * S_1 * S_2\,,\, T_1 * S_2 + S_1 * T_2)$$

associative owing to the fact that λ is a central element of $\mathcal{L}$. *Endowed with this product* $\tilde{\mathcal{L}}$ *becomes a complex algebra containing* $\mathcal{L}$ *as a subalgebra with embedding*

$$(7,13) \qquad \mathcal{L} \ni T \rightarrow (T,0) \in \tilde{\mathcal{L}}$$

and containing $\lambda^{\frac{1}{2}}$ *as* $\left(0,\delta_0\,\lambda^{\frac{1}{2}}\right)$ *with square* $\lambda = (\delta'_0 1 0)$.

The rest of our program now unfolds as follows : *one gets the trace* $\mathcal{T}$ of $\tilde{\mathcal{L}}$ *as*[31]

$$(7,14) \qquad \mathcal{T}((T,S)) = \sqrt{\pi}\ \mathrm{Tr}\,(S(1)), \qquad , (T,S)\in \tilde{\mathcal{L}}\,,$$

the homomorphism $\pi_0 : A \rightarrow \tilde{\mathcal{L}}$ *as*

$$(7,15) \qquad \pi_0(a) = ((a)\delta_0,0) = (a,0),\ a\in A\ ;$$

[30] Note that the map $a \rightarrow \delta_0(a)$, $a\in A$, δ_0 the Dirac measure carried by $\{0\}$, then yields a ∗-homomorphism : $A \rightarrow \mathcal{L}$.

[31] We shall use the shorthand $(a)\delta_0 = a$, $a\in A$.

and furthermore the involution $F \in \tilde{\mathcal{L}}$ through a construction which we now describe. Setting

$$(7,16) \qquad N(f) = \frac{1}{\sqrt{\pi}} \int_{\infty}^{0} f(s)\ s^{-\frac{1}{2}}\ e^{-sD^2}\ ds\ , \quad f \in \mathcal{S},$$

is a convergent integral which defines an element N of $\mathcal{L}^1$ with Laplace transform

$$(7,17) \qquad N^L(p) = \frac{1}{\sqrt{D^2+p}}$$

Moreover, there is a unique element of $\mathcal{L}$, noted DN, with Laplace transform

$$(7,18) \qquad DN^L(p) = \frac{D}{\sqrt{D^2+p}}$$

and whose corresponding complex operator-valued function

$$(7,19) \qquad DN(z) = \frac{1}{\sqrt{\pi}} \frac{1}{\sqrt{z}} De^{-zD^2}$$

is holomorphic for s = Rez > 0 and such that

$$(7,20) \qquad \sup_{z \in \frac{1}{p} U_r} \|DN\| \leq \frac{p}{1-r} \|De^{-sD^2}\| \quad , \quad r > 0$$

Defining

$$(7,21) \qquad F = (DN, \varepsilon N)$$

then yields an element of $\mathcal{L}$ *of unit square*. It then follows from (7,1) above that , ϕ *defined for* $a_0, a_1, \ldots a_{2n} \in A$, *as*

$$(7,22) \qquad \begin{cases} \phi(a_0 da_1 \ldots da_{2n}) = \mathcal{T}\{Fa_0[F,a_1] \ldots [F,a_{2n}]\} \\ \phi(a_0) = \mathcal{T}(Fa_0) \end{cases}$$

is a normalized cyclic cocycle of A, which is in addition entire[32] . ϕ is called the **character of the θ–summable Fredholm module** ($\mathcal{H}$,D). Note that one has, in zero $\mathbb{N}$-grade

$$(7,23) \qquad \phi(a_0) = T_r\left\{\varepsilon(a_0)\, e^{-D^2}\right\} \quad , a_0 \in A.$$

We close up this section by mentioning an expression of the character ϕ as an integral over the mass : we have, for each real $\alpha > 0$

$$(7,24) \qquad \phi(a_0 da_1 \ldots da_{2n}) = \frac{1}{\pi}\, \Gamma\left(n + \frac{1}{2}\right)\, e^{(im+\alpha)^2)}$$

$$\int_{-\infty}^{+\infty} F(im + \alpha)\ \mathbf{a}_0 \left[F(im+\alpha),\mathbf{a}_1\right] \ldots \left[F(im/\alpha),\, \mathbf{a}_{2n}\right] dm$$

where

$$(7,25) \qquad F(z) = \frac{D+z\varepsilon}{\sqrt{D^2+z^2}} \quad , \quad z \in \mathbb{C}.$$

[32] cf. section 5 of [1] for a proof.

APPENDIX A. A SKETCH OF ORDINARY CYCLIC COHOMOLOGY (OF $\mathbb{Z}/2$-GRADED COMPLEX ALGEBRAS)

For the benefit of the reader using this report as an introduction, we now include an outline of ordinary cyclic cohomology (intended as a leading thread through [1] and [7]). The technical perspective is that of [7] (the reader may also consult [8], [8a] : for an historical introduction, see [9] and the introduction of [7][33]).

In this Appendix $A = A^0 \oplus A^1$ is a $\mathbb{Z}/2$-graded complex algebra. The **Hochschild cohomology of** A (with values in the algebraic dual of A considered as an A-bimodule) is obtained as follows in terms of linear forms of $\Omega(A)$:

The **Hochschild cochains** are the closed linear forms of $\Omega(A)$, whose set $C = \bigoplus_{n\in N} C^n$ was introduced in § 2 (cf. (2,1)).

The **Hochschild boundary operator** $b : C \to C$, is then defined as the transpose of $\beta\varepsilon$ [34] :

$$b\varphi = \varphi \cdot \beta\varepsilon \quad , \quad \varphi \in C, \tag{A,1}$$

with the following sets $Z, B \in C$ of **Hochschild cocycles resp. boundaries**

$$Z = \mathrm{Ker}\, b \quad , \quad B = \mathrm{Im}\, b, \tag{A,2}$$

and with the **Hochschild cohomology group**

$$H = Z/B = \mathrm{Ker}\, b/\mathrm{Im}\, b \tag{A,3}$$

($H = \bigoplus_{n\in N} H^n$ with $H^n = Z^n/B^n$, $Z^n = Z \cap C^n$, $B^n = B \cap C^n$). Interpreting the C^n as (n+1)-multilinear forms on A through the convention[35]

[33] The exposition of cyclic cohomology sketched here follows the computational approach of [1]. The reader interested in homological algebra aspects may consult [13] through [19].

[34] One has $b^2 = 0$ (cf. (1,37)

[35] cf.[7] Remark [3.5] and Appendix K ; and the last section of [12]

$$\text{(A,4)}\quad \varphi(a_0,a_1,\ldots,a_n) = (-1)^{n\sum_{k=0}^{n}\partial a_k + \sum_{k\,\mathrm{odd}}\partial a_k}\ \varphi(a_0 da_1 \ldots da_n),\ a_0,\ldots,\ a_n \in A^0 \cup A^1$$

b then reads[36]

$$\text{(A,5)}\qquad (b\varphi)(a_0,\ldots,a_{n+1}) = \sum_{i=0}^{n} (-1)^i\, \varphi(a_0,\ldots,a_i a_{i+1},\ldots,a_{n+1})$$

$$- \ (-1)^{n+\partial a_{n+1}\sum_{k=0}^{n}\partial a_k}\ \varphi(a_{n+1}a_0,a_1,\ldots,a_n)$$

The **cyclic cohomology** of A now arises as the cohomology of the **cyclic subcomplex** (C_λ, b) of the Hochschild complex (C,b) : where C_λ is defined as the set of **cyclic linear forms** of A (i.e. such that $\varphi\circ\lambda = \varphi$[37] ; one has $C_\lambda \subset C$ owing to (1,20)). The fact that (C_λ, b) is a subcomplex follows from the following Lemma, itself an immediate consequence of relation (1.26) :

[A.1] Lemma. *Let ϕ be a linear form of $\Omega(A)$. Then*

(i) ϕ is a graded trace of $\Omega(A)$ iff $\phi\circ\beta = \phi\circ\gamma = 0$

(ii) *ϕ is cyclic iff $\phi\circ d = \phi\circ\gamma = 0$*

(iii) *ϕ is a closed graded trace of $\Omega(A)$ iff $\phi\circ d = \phi\circ\beta = 0$*

(iv) *if ϕ is cyclic , so is $\phi\circ\beta$*

We shall denote by Z_λ, resp. B_λ, the **cyclic cocycles** resp. **cyclic boundaries** of A and by $H_\lambda = Z_\lambda/B_\lambda$ its **cyclic cohomology group** (with $C^n_\lambda = C_\lambda \cap C^n$, $Z^n_\lambda = Z_\lambda \cap C^n$, $B^n_\lambda = B_\lambda \cap C^n$ and $H^n_\lambda = Z^n_\lambda/B^n_\lambda$). Summarizing, we have

---

[36] $\sum_{i=0}^{n}$ in the r.h.s. of (A,5) corresponds to the transverse of β'ε, and the last ("flip over") term to that of α. For a trivially graded, (A,5) yields the usual definition of Hochschild cohomology with coefficients in the dual.

[37] One has $(\varphi\circ\lambda)(a_0,\ldots,a_n) = (-1)^{n+\partial a_n \sum_{k=0}^{n-1}\partial a_k}\ \varphi(a_n,a_0,\ldots,a_{n-1})$

(A,6) $$C_\lambda = \left\{\phi\in \Omega(A)^* \,;\, \phi = \phi 0\lambda\right\} = \left\{\phi\in \Omega(A)^* \,,\, \phi 0d = \phi 0\gamma = 0\right\}$$

(A,7) $$\begin{aligned} Z_\lambda &= \left\{\phi\in \Omega(A)^* \,;\, \phi 0\lambda = \phi \text{ and } b\phi = 0\right\} \\ &= \left\{\phi\in \Omega(A)^* \,;\, \phi 0d = \phi 0\beta = 0\right\} \\ &= \text{set of closed graded traces of } \Omega(A) \end{aligned}$$

(A,8) $$B_\lambda = \left\{b\phi \,;\, \phi\in C_\lambda\right\}$$

As the simplest example of cyclic cohomology, let us describe that of the ground field $\mathbb{C}$ itself. The differential envelope $\Omega(\mathbb{C})$ has a linear basis consisting of : 1, $1(d1)^n$ and $(d1)^n$, $n = 1,2,\ldots$. The Hochschild cochains (closed linear forms) $\phi\in C(\mathbb{C})$ are one-to-one with the numbers

(A,9) $$C_n(\phi) = \phi(1(d1)^n)$$

One easily computes that

(A,10) $$\begin{cases} Z^\lambda(\mathbb{C}) = C^\lambda(\mathbb{C}) = \left\{\phi\in C(\mathbb{C}) \,;\, C_n(\phi) = 0 \text{ for all odd } n\in \mathbb{N}\right\} \\ B^\lambda(\mathbb{C}) = \{0\} \end{cases}$$

so that *one has*

(A,11) $$Z^n_\lambda(\mathbb{C}) = H^n_\lambda(\mathbb{C}) = \begin{cases} 0, & n \text{ odd} \\ \mathbb{C}, & n \text{ even} \end{cases} .$$

In addition to its linear structure, cyclic cohomology has a module structure for the cup product # defined as follow. Given $\phi\in \Omega(A)^*$ and $\psi\in \Omega(B)^*$, A, B two $\mathbb{Z}/2$-graded complex algebras, their **cup product** $\phi\#\psi$ is the element of $\Omega(A\otimes B)^*$ given by

(A,12) $$\phi\#\psi = (\phi\otimes\psi)\, o\, \theta$$

where $\theta\in \mathrm{Hom}(\Omega(A\otimes B), \Omega(A)\otimes\Omega(B))$ is defined via universality of $\Omega(A\otimes B)$ (cf (1,10)) :

(A,10)

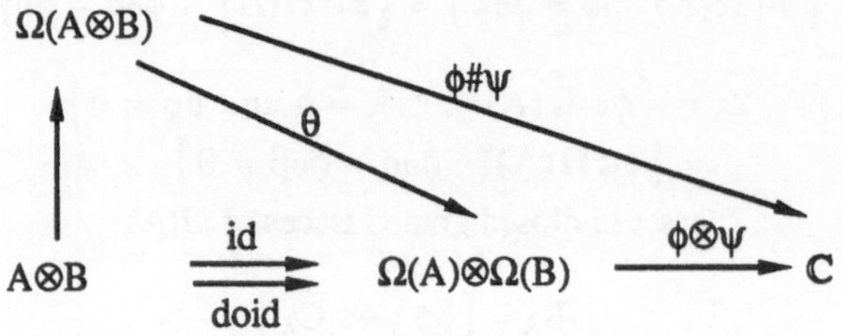

(here A ⊗ B is a skew product of $\mathbb{Z}/2$-graded complex algebras, $\phi \otimes \psi$ an ordinary tensor product of linear forms, and $\Omega(A) \otimes \Omega(B)$ a skew product of $\mathbb{N}/2$-graded differential algebras (see [7] for details) so that $\left\{\Omega(A) \otimes \Omega(A)\right\}^{\circ} \cong A \otimes B$). *The cup product* $\phi\#\psi$ *of two cyclic cocycles* $\phi \in Z^n_\lambda(A)$ *and* $\psi \in Z^n_\lambda(B)$ *is a cyclic cocycle* $Z^{n+m}_\lambda(A \otimes B)$ (as the graded trace of $\Omega(A \otimes B)$, pull back of the graded trace $\phi \otimes \psi$ of $\Omega(A) \otimes \Omega(B)$ by the homomorphism θ of $\mathbb{Z}/2$-graded algebras). If ϕ *or* ψ *is in addition a cyclic boundary, so is* $\phi \# \psi$: thus *the cup product passes to the cohomology groups which thereby acquire a module structure*. For instance $H_\lambda(\mathbb{C})$ is, as a module for the cup product, generated in order 2 by the cohomology class of the cyclic 2-cocycle $\sigma_0 \in Z^2_\lambda(\mathbb{C})$ defined by

(A,11) $$\sigma_0(1d1d1) = 2\pi i$$

Effecting the cup product by σ_0 to the right

(A,12) $$\phi \to \sigma_0 \# \phi \quad , \phi \in \Omega(A)^*$$

one gets, owing to $A \otimes \mathbb{C} \cong A$ a grade-two linear operator of $\omega(A)$ yielding **Connes'operator** S on cyclic cocycles :

(A,13) $$S\tau = \sigma_0 \# \tau \; (= \tau\#\sigma) \quad , \tau \in Z_\lambda(A)$$

(S *passes to the cohomology groups* ; and is extended to cyclic chains by setting

(A,13a) $$S\phi = \frac{1}{n+3}(\sigma_0 \# \phi) \quad , \phi \in C^n_\lambda(A))$$

For the computation of S we use as follows the "calculus within $\Omega(A)$" : *we have*

(A,14) $$\sigma_0 \# \phi = 2\,\pi i\, \varphi \circ \Sigma \quad , \phi \in \Omega(A)^*$$

where the operator Σ *of* $\Omega(A)$ *(of grade-2) reads as follows* in terms of the operators of section 1[38] :

(A,15) $$\Sigma = \beta(\xi-\alpha) + (\xi+\alpha\rho^{-1})\beta$$

Applying this to $Z_\lambda(A)$ now yields

[A.2] Proposition *with* S *given by* (A,13) *we have that*
(i) S *turns cyclic cocycles into Hochschild boundaries : for* $\tau \in Z_\lambda^{n-1}$, *we have that*

(A,16) $$\frac{1}{2\pi i} S\tau = n(n+1)\, b\, \psi_0 \; (=n(n+1)\, b\, \psi_0\beta\varepsilon)$$

where $\psi_0 \in C^n(A)$ *is given by*

(A,17) $$\psi_0 = \frac{\tau\xi\varepsilon}{n(n+1)}$$

(ii) *the failure of* ψ_0 *to be cyclic is indicated by*

(A,18) $$\tau\xi\varepsilon(1-\lambda) = (n+1)\,\tau\alpha \; (= (n+1)\tau\varepsilon\alpha = (n+1)\tau\beta')$$

(iii) *one gets* τ *back from* ψ_0 *by applying the operator* B :

(A,19) $$\tau = B\psi_0 \; (= \psi_0 \mathbb{B} = n\psi_0 \mathbb{B}_0 = nB_0\psi_0)$$

$$b\psi_o\varepsilon \in Z_\lambda^{n+1}$$

$$b$$

(A,20) $$C^n \ni \psi_0 = \frac{\tau\xi\varepsilon}{n(n+1)} \qquad S/2\pi i n(n+1)$$

$$B$$

$$\tau = B\psi_0 \in Z_\lambda^{n-1}$$

---

38 see [7] p. 67 for a proof

Indications on the proof : the exploration of the situation (A,20) in fact furnishes the motivation for defining the operator B. When we compute $S\tau = \sigma_0\#\tau = 2\pi i\tau 0\Sigma$ for $\tau(=\tau\varepsilon) \in Z_\lambda^{n-1}$, we find that only the third term in the sum (A,15) survives : indeed, we have, on the one hand $\tau\beta = 0$ by definition, and on the other (replacing Σ by $\Sigma\varepsilon$ as we may since $\tau = \tau\varepsilon$), we have $\tau\alpha\rho^{-1}\beta\varepsilon = \tau\alpha\rho^{-1}\varepsilon\beta\varepsilon = 0$ by (1,36) and (1,38) : we thus get (A,16) with (A,17). We then naturally compute $\psi_0(1-\lambda)$, a somewhat lengthy calculation[39] leading to (A,18). Now the fact that α has the right inverse σ in restriction to $A\Omega(A)$ (cf. (1,39)) suggests the introduction of $\mathbb{B}_0$ and $\mathbb{B}$ (cf. (1,29) through (1,31)) allowing to recover τ from ψ_0 as in (A,19)). The properties (1,32) and (1,33) of $\mathbb{B}$ entail the relations $B^2 = 0$ and $Bb = -bB$ quoted in (2,6) (properties playing a basic role in the definition of the entire cyclic homology). On the other hand one has evidently $C_\lambda \subset \operatorname{Ker} B$ and $\operatorname{Im} B \subset C_\lambda$, in fact *the range of* B *covers* C_λ [40] :

(A,21) $$C_\lambda = \operatorname{Im} B$$

The role of B (the most important operator of C besides b and S) in ordinary cyclic cohomology is the link which it creates between cyclic and Hochschild cohomology (crucial for computing the latter) : indeed

[A.3] Theorem *One has a long exact sequence*

(A,22) $$\rightarrow H^n \xrightarrow{B_*} H_\lambda^{n-1} \xrightarrow{S_*} H_\lambda^{n+1} \xrightarrow{I_*} H^{n+1} \longrightarrow$$

where

[39] see [7] p.70

[40] see [7] p.74, Sub lemma [6.3]

(A,23)
$$\begin{cases} B_*\,(\xi \bmod B(A)) = B\xi \bmod B_\lambda(A) & ,\ \xi \in H \\ S_*\,(\tau \bmod B_\lambda(A)) = S\tau \bmod B_\lambda(A) & ,\ \tau \in Z_\lambda \\ I_*\,(\tau \bmod B_\lambda(A)) = \tau \bmod B(A) & \end{cases},$$

in other terms, one has an exact triangle

(A,22a)
$$\begin{matrix} & H & \\ B_* & & I_* \\ & S_* & \\ H_\lambda & & H_\lambda \end{matrix}$$

(in fact (A,22) *is identical with the cohomology long exact sequence of the short exact sequence* $0 \to C_\lambda \to C \to C/C_\lambda \to 0$) (where S_* and I_* are induced in cohomology by S and the natural injection $Z_\lambda^* \subset Z$). We now show how one comes to the scheme (A,22), (A,23). In view of property (A,21) the situation (A,20) raises the following question : to which extent is the diagram (A,20) commutative ? More precisely :

(*) *Given the cyclic cocycle* $\tau \in Z_\lambda^{n-1}$, *with* $\psi \in \overset{-1}{B}(\tau) \subset C^n$ *(defined up to* Ker B, *what is the relation of* $b\psi$ *(defined up to* b(Ker B)) *to* τ ?

This question will lead us to a crucial enlargement of the playground, linked with the following square of linear spaces

(A,23)
$$\begin{cases} Z_\lambda = \operatorname{Ker} b \cap \operatorname{Im} B \subset \operatorname{Ker} b \cap \operatorname{Ker} B \ (\text{say } Z_\pi) \\ \qquad\quad \cup \qquad\qquad\qquad\qquad \cup \\ B_\lambda = b(\operatorname{Im} b) \qquad \subset b(\operatorname{Ker} B) \quad (\text{say } B_\pi) \end{cases}$$

where the indicated inclusions allow to define an obvious linear map $\kappa : H_\lambda = Z_\lambda/B_\lambda \to \frac{Z_\pi}{B_\pi}$. A crux of the theory is now

[A.4] Proposition[41] *The canonical map*

$$\kappa : H_\lambda = Z_\lambda/B_\lambda \to \text{Ker b} \cap \text{Ker B}/\text{b}(\text{Ker B}) \tag{A,24}$$

is a linear bijection (injective : $Z_\lambda \cap B_\pi = B_\lambda$ *and surjective* : $Z_\lambda + B_\pi = Z_\pi$). *Consequently cyclic cohomology can be considered as defined by the space of cocycles* $Z_\pi = \text{Ker b} \cap \text{Ker B}$ *modulo the space of boundaries* $B_\pi = \text{b}(\text{Ker B})$[42].
The bijection κ allows to answer the question (∗) in the following way.

[A.5] Proposition. Let $\tau \in Z_\lambda^{n-1}$, $n \geq 1$. *Interpreting* $\text{b}\overset{-1}{\text{B}}(\tau)$ *as an element of* Z_π/B_π *we have that*

$$\kappa^{-1}\text{b}\overset{-1}{\text{B}}(\tau) = \frac{1}{2\pi i n(n+1)} S_*[\tau] \tag{A,25}$$

In other terms $S_*[\tau]$ *is the class in* H_λ^{n+1} *of any* $\tau' \in Z_\lambda^{n+1}$ *such that* $\tau = B\psi$, $\psi \in C^n$, $\tau' - b\psi \in B_\pi$.

Our next item is new the *double complex* C^{**} :

$$\begin{array}{ccccccc} 0 & \to & 0 & \to & C^0 & & C^1 \\ \uparrow & & \uparrow & & \uparrow & & \uparrow \\ 0 & \to & C^0 & & C^1 & & C^2 \\ \uparrow & & \uparrow d_2 & & \uparrow & & \uparrow \\ C^0 & \to & & & C^1 & \to & C^2 \to C^3 \\ & d_1 & & & & & \end{array} \tag{A,26}$$

with horizontal, resp. vertical, and total differentials d_1, resp. d_2, and Δ (cf. (2,7), (2,8)). Note that, since the parallels to the first diagonal are constant, the *shift* s *by* 2 *parallel to the second diagonal* :

---

41 This is Lemma [6.6] of [7], transposition to the **Z/2**-graded case of Lemma 36 of [0].

42 The role of the cocycles in Z versus those in Z_λ in cyclic cohomology parallels that of the general versus the normalized cocyles in entire cyclic cohomology. This will become clearer after Proposition [A.6] below.

(A,27) $$s\phi^{n,m} = \phi^{n+2,m+2} \quad , \quad (\phi^{nm}) \in C^{**}$$

is a homomorphism of C^{**} :

(A,28) $$\begin{cases} sd_1 = d_1 s \\ sd_2 = d_2 s \end{cases}$$

Denoting by ${}^{(1)}E_r$, resp. ${}^{(2)}E_r$ the first, resp. second spectral sequence of C^{**}, corresponding to the standard filtrations

(A,29) $$\begin{cases} {}^{(1)}F_p C^{**} = \bigoplus_{n \geq p} C^{n,m} \\ {}^{(2)}F_q C^{**} = \bigoplus_{n \geq q} C^{n,m} \end{cases}$$

we then have

[A.6] Proposition

(i) *The first spectral sequence* of C^{**} *is degenerate*

(A,30) $${}^{(1)}E_2^{p,q} = 0 \quad , q \neq 0$$

(ii)*The cohomology of the total complex of* C^{**} *coïncides with the cyclic cohomology of* A

(A,31) $$H^n(\mathrm{Tot}C^{**}) = H^n(C^* \cap \mathrm{Ker}\, B, b) = H_\lambda^n$$

(iii)*The operator* s_* *of grade 2 induced by* s *on* $H^*(\mathrm{Tot}C^{**})$ *coïncides with* $\frac{i}{2\pi} S_*$ *via the above isomorphism* $H^*(\mathrm{Tot}C^{**}) = H_\lambda^*$

(iv) *The exact couple determined by the second filtration of* C^{**} *coïncides with the exact couple* (A,22a).

The proof is straightforward, granted the above results (we refer to [7] pp. 87 and following). The equality (A,31) is obtained from the familiar "zick-zack process" (cf. [17] Chapter III). Observe that elements of Z_π = Ker b $\cap$ Ker B are (finite) entire cocycles, and that, *for a* $\tau \in Z_\pi \in$ Ker B *to be normalized in the sense of section 4, it is necessary and sufficient that it be an element of* Z_λ
(cf. Remark [A.12] below, and footnote 43 above).

We end up this section with a sketch of the techniques of proof. Whilst the preceding [A.6] through [A.6] are the central results of the theory, the basic technical tools are the three following (rather opaque)[43].

[A.7] Lemma *Let* $\tau \in Z_\lambda^{n-1}$ *with* ψ_0 *as in* (A,17).
For any $\psi \in B^{-1}(\tau)$ *we have*

(A,32) $$(\psi-\psi_0)(1-\lambda) = \psi\beta\varepsilon\, B_0 + \psi\, B_0\, \Lambda^\perp \beta'\varepsilon$$

i.e. choosing a $\varphi \in C^{n-1}$ *such that* $\psi B_0\, \Lambda^\perp = \varphi(1-\lambda)$:

(A,33) $$(\psi-\psi_0-\varphi\beta\varepsilon)(1-\lambda) = \psi\beta\varepsilon\, B_0$$

[A.8] Lemma *Let* $\psi \in C^n$ *be such that*

(A,34) $$\psi B_0 \Lambda^\perp = \varphi(1-\lambda) \quad , \quad \varphi \in C^{n-1},$$

we have that

(A,35) $$\varphi B_0 \beta'\varepsilon (= \psi B_0\, \Lambda^\perp - \varphi\beta\varepsilon B_0) = (\psi-\varphi\beta\varepsilon)\, B_0 - \frac{1}{n+1}\psi\, B = \psi\beta\varepsilon\, B_0(1-AN^{-1})\sigma$$

[A.9] Lemma *Let* $\varphi \varepsilon C$ *be such that* $\varphi B_0 \beta'\varepsilon = 0$.
Then $\varphi\beta\varepsilon \in$ b(Ker B)

---

43 [A.6] is the quintessence of the proof of Lemma [6.5] of [7], transposition of Lemma 34 of [0]. [A.7] is essentially Lemma] [6.9] of [7] slightly streched so as to accomodate the proof of Proposition [3.1].

Proofs of these Lemmas : [A.6] : we have (cf. (A.18))

(A,36) $$\psi_0(1\text{-}\lambda) = \frac{1}{n}\tau\beta' = \frac{1}{n}\psi\mathbb{B}\beta'$$

thus, by (1,40)

(A,37) $$\psi(1\text{-}\lambda) = \psi(1\text{-}\lambda)\varepsilon = \psi(\beta\mathbb{B}_0 + \mathbb{B}_0\beta')\varepsilon = \psi\beta\varepsilon\mathbb{B}_0 + \psi\mathbb{B}_0\beta'\varepsilon$$

hence

(A,38) $$(\psi\text{-}\psi_0)\,(1\text{-}\lambda) = \psi\beta\varepsilon\mathbb{B}_0 + \psi\mathbb{B}_0(1 - \frac{1}{n}A)\,\beta'\varepsilon$$

proving (A,32). With φ as indicated we have, using (1,41)

(A,39) $$\psi\mathbb{B}_o\,A^{\perp}\,\beta'e = \varphi(1\text{-}\lambda)\,\beta'\varepsilon = \varphi\beta\varepsilon(1\text{-}\lambda)$$

whence (A,33). For the proofs of [A.8] and [A.9] we refer to [6], [1.6] and [1.7].

Utilisation of these Lemmas

As a corollary of Lemma [A.7], we have[44]

[A.10] Lemma *Let* $\psi\varepsilon C^n$ *be such that* $b\psi\in C_\lambda^{n+1}$. *Then*

(i) $B\psi = \tau\in Z_\lambda^{n-1}$

(ii) $b\psi\in Z_\lambda^{n+1}$

(iii) $S\tau - 2\pi in\,(n+1)\,b\psi\in B_\lambda^{n+1}$

hence we have $S_*[\tau] = 2\pi in\,(n+1)\,[b\psi]$

[44] [A.10] (cf. Lemma [6.5] of [7], transcription of Lemma of [0]) gives, in the special case when $b\psi$ is cyclic, an answer to question (*) within the original frame of definition of cyclic cohomology.

Proof : (i) : $\tau \in \operatorname{Im} B = C_\lambda$ and $b\tau = bB\tau = -Bb\tau = 0$; (ii) $b\psi \in B^{n+1} \cap C_\lambda^{n+1} \subset Z_\lambda^{n+1}$; since $S\tau = 2\pi i n\,(n+1)\, b\psi_0$, (iii) boils down to $b(\psi-\psi_0) \in B_\lambda^{n+1}$: however this follows from (A,33) whose r.h.s. vanishes.

The preceding Lemma now has the following

[A.11] Corollary *We have that*

(i) $SBZ^n \subset B_\lambda^{n+1}$, $n \in \mathbb{N}$

(ii) $\operatorname{Im} B \cap b(\operatorname{Ker} B) = b(\operatorname{Im} B)$ $(\Rightarrow Z_\lambda \cap B_\pi = B_\lambda)$

(iii) $\operatorname{Im} S_* = \operatorname{Ker} I_*$

(note that (ii) is the injectivity result in Proposition [A.4], and that (vi) is exactness of the triangle (A,22a) in H_λ.

Proof:

(i) for $\psi \in Z^n$, we have $B\psi = \tau \in BZ^n$ and $b_\psi = 0$, thus [A.10] (iii) yields $S\tau = SB\psi \in B_\lambda^{n+1}$

(ii) for $\psi \in$ Ker B, we have $\tau = B\psi = 0$, thus if moreover $b\psi \in C_\lambda = \operatorname{Im} B$, [A.10] (iii) yields $b\psi \in B_\lambda = b\,(\operatorname{Im} B)$

(iii) $\subset$ follows from [A.2] (i); $\supset$: for $\tau_1 \in Z_\lambda^{n+1}$ one has $[\tau_1] \in \operatorname{Ker} I_*$ iff $\tau_1 \in \operatorname{Im} B$, [A.10] then yielding $SB\psi - 2\,\pi i n(n+1)\,\tau_1 \in B_\lambda$ i.e. $[\tau_1] \in \operatorname{Im} S_*$

We now interpose

[A.12] **Remark** *We have that* $\operatorname{Ker} b \cap \operatorname{Ker} B_0 = Z_\lambda$, *in other terms* $\operatorname{Ker} B_0 \cap Z_\pi = Z_\lambda$. Consequently $Z_\nu \cap Z_\pi = Z_\lambda$.

Proof : $\supset$: we have $Z_\lambda \subset \operatorname{Ker} b$ and $Z_\lambda \subset C_\lambda \subset \operatorname{Ker} B_0$; $\subset$: follows from $(1-\lambda)\varepsilon = \beta\varepsilon\, \mathbb{B}_0 + \mathbb{B}_0\, \beta'\varepsilon$ (cf. (1,40), (1,42)). Now, $\xi \in Z_\nu \cap Z_\pi$ is s.t. $\xi\, \mathbb{B}_0 = \xi\, \mathbb{B}, (N+1)^{-1} = 0$.

We finally show the surjectivity part of Proposition [A.5] (i.e. $Z_\pi = Z_\lambda + B_\pi$) as a consequence of Lemmas [A.8] and [A.9] : applying the last equation (A,35) to $\psi \in Z_\pi = \operatorname{Ker} b \cap \operatorname{Ker} B$ we see that $(\psi - \varphi\beta\varepsilon)\, \mathbb{B}_0 = 0$, hence, using Remark [A.12]

$\psi - \varphi\beta\varepsilon \in \operatorname{Ker} b \cap \operatorname{Ker} B_0 = Z_\lambda$. However, since $\varphi\, \mathbb{B}_0\, \beta'\varepsilon = 0$, again by (A,35), Lemma [A.9] yields $\varphi\beta\varepsilon \in b\,(\operatorname{Ker} B) = B_\pi$.

We already showed exactness of the triangle (A,22a) in the right vertex. The two other exactness conditions are easily checked using the above stated result (see [7]. 80).

APPENDIX B. THE CUNTZ DEFORMATION $(Q_t\Omega,\tau,\underset{t}{*},d)$ AND ZEKRI ALGEBRA $(\mathfrak{E}_tA,\hat{\tau},\underset{t}{*},\tau)$ of a $\mathbb{Z}/2$-graded differential algebra (Ω,θ,d)

[B.1] Proposition. *Let* $(\Omega, \Omega^+ \oplus \Omega^-, d)$ *be a* $\mathbb{Z}/2$-graded *differential k-algebra* ($k = \mathbb{R}$ or $\mathbb{C}$) *and let* $t \in k$. *We denote by* θ *the grading automorphism of* Ω *and by* $\partial\omega \in \mathbb{Z}/2$ *the grade of*
$\omega \in \Omega^+ \cup \Omega^-$[45]. *Then*

(i) *the bilinear product* $\underset{t}{*}$ *given by*

$$\text{(B.1)}\quad \omega \underset{t}{*} \omega' = \begin{cases} \omega\omega' & , \omega \in \Omega^+ \\ \omega\omega' + t\omega d\omega' & , \omega \in \Omega^- \end{cases}$$

is associative, and such that

$$\text{(B.2)}\quad \begin{cases} d(\omega \underset{t}{*} \omega') = (d\omega)\underset{t}{*}\omega' + (-1)^{\partial\omega}\omega \underset{t}{*} d\omega' - t(-1)^{\partial\omega}(d\omega)\underset{t}{*}(d\omega') \\ \omega \in \Omega^{(\partial\omega)},\ \omega' \in \Omega \end{cases}$$

(ii) *the linear map* $\tau : \Omega \to \Omega$ *given by*

$$\text{(B.3)}\quad \tau\omega = (-1)^{\partial\omega}(\omega - td\omega) \quad , \Omega^+ \cup \Omega^-$$

i.e.

$$\text{(B.3a)}\quad \tau = (1-td)\,\theta = \theta(1+td)$$

is an involutive automorphism of Ω *anticommuting with* d :

$$\text{(B.4)}\quad \tau(\omega \underset{t}{*} \omega') = (\tau\omega) \underset{t}{*} (\tau\omega') \quad , \omega,\omega' \in \Omega$$

$$\text{(B.5)}\quad \tau^2 = \mathrm{id}$$

$$\text{(B.6)}\quad \tau d = -d\tau$$

---

[45] In other words : $= \Omega^+ \oplus \Omega^-$ is an associative k-algebra with an automorphism θ s.t., for $\omega \in \Omega^+$, $\theta\omega = (-1)^{\partial\omega}\omega = \pm\omega$ (hence $\theta^2 = \mathrm{id}$) ; and with a graded derivation d such that $d^2 = 0$ and $d\theta = -\theta d$.

Note that, using τ, *property* (B,2) *reads*

(B.2a) $$d(\omega \underset{t}{*} \omega') = (d\omega) \underset{t}{*} \omega' + (\tau\omega) \underset{t}{*} d\omega' \ , \ \omega, \omega' \in \Omega$$

so that (i) *and* (ii) *assert that, under the product* $\underset{t}{*}$, *the grading involution* τ, *and the differential* d, Ω *is a* $\mathbb{Z}/2$*-graded algebra which we henceforth denote* $Q_t\Omega$.

[B.2] Proposition. *With* Ω *and* $Q_t\Omega$ *as in* [B.1], *and* $\alpha : \mathbb{Z}/2 \to \mathrm{Aut}\, Q_t\Omega$ *the group homomorphism*

(B.7) $$\begin{cases} \alpha(0) = \mathrm{id} \\ \alpha(1) = \tau \end{cases}$$

we denote by $\mathcal{E}_t\Omega$ *the crossproduct*

(B.8) $$\mathcal{E}_t\Omega = Q_t\Omega \underset{\alpha}{\times} \mathbb{Z}/2 \ ,$$

and by $\hat{\alpha} : \widehat{\mathbb{Z}/2} = \mathbb{Z}/2 \to \mathrm{Aut}\, \mathcal{E}_t\Omega$ *the automorphism group dual of* α. $\mathcal{E}_t\Omega$ *is the* $\mathbb{Z}/2$–graded *algebra obtained as the vector space*

(B.9) $$\mathcal{E}_t\Omega = Q_t\Omega \oplus Q_t\Omega$$

with the bilinear associative product :

(B.10) $$\begin{cases} (\omega_0,\omega_1)\underset{t}{*}(\omega'_0,\omega'_1) = (\omega_0\underset{t}{*}\omega'_0+\omega_1\underset{t}{*}\tau\omega'_1,\omega_0\underset{t}{*}\omega'_1+\omega_1\underset{t}{*}\tau\omega'_0) \\ (\omega_0,\omega_1),(\omega'_0,\omega'_1)\in \mathcal{E}_t\Omega \end{cases}$$

and the grading involution $\hat{\tau} = \hat{\alpha}(1)$ *given by*

(B.11) $$\hat{\tau}((\omega_0,\omega_1)) = (\omega_0,-\omega_1) \quad , \quad (\omega_0,\omega_1)\in \mathcal{E}_t\Omega.$$

Under the map

(B.12) $$(\omega_0,\omega_1) \to \begin{pmatrix} \omega_0 & \omega_1 \\ \tau\omega_1 & \tau\omega_0 \end{pmatrix}$$

$\mathcal{E}_t\Omega$ *appears as a* $\mathbb{Z}/2$-graded *subalgebra of* $M_2(Q_t\Omega) = M_2(\mathbb{C}) \otimes Q_t\Omega$ *endowed with the grading automorphism* $\varepsilon \otimes \mathrm{id}$, $\varepsilon = \mathrm{ad}\begin{pmatrix}1 & 0\\ 0 & -1\end{pmatrix}$.

Note also that one gets another involutory automorphism τ *of* $\mathcal{E}_t\Omega$ *commuting with* $\hat{\tau}$ *by setting* :

$$\tau((\omega_o, \omega_1)) = (\tau\omega_0, \tau\omega_1)\ ,\ (\omega_0, \omega_1) \in \mathcal{E}_t\Omega. \tag{B.13}$$

The proofs of **[B.1]** and **[B.2]** are straightforward verifications.

[B.3] Remarks. (i) *If* Ω *has a unit* 1, *this unit (automatically of grade 0, and vanishing under* d) *is also the unit of* $Q_t\Omega$, $t \in k$. *On the other hand the process* $\Omega \to \tilde{\Omega} = \mathbb{C}\,\tilde{1} \oplus \Omega$ *of adding a unit* $\tilde{1}$ *commutes with* $Q_t : Q_t\tilde{\Omega} = (Q_t\Omega)^{\sim}$.

(ii) *Note that* $\mathcal{E}_t\Omega$ *can be considered as consisting of the formal sums* $\omega_0 \oplus \omega_1 f$, $\omega_0\omega_1 \in \Omega$, *with* $f^2 = 1$ *and* $f\omega f = \tau\omega$, $\omega \in \Omega$. *If* Ω *has a unit* 1, $f = (0,1) \in \mathcal{E}_t\Omega$; *if not*, $f \in \mathcal{E}_t\tilde{\Omega}$ *as* $(0, \tilde{1})$.

(iii) *Assuming that* Ω *has a unit* 1, *and with* B *a unital algebra with unit* 1_B, *there is a bijection between*

(a) *the unital homomorphisms* $\pi : E_t\Omega \to B$ of complex algebras

(b) *the pairs* (π, f) *of a unital homomorphism* $Q_t\Omega \to B$ *of complex algebras and of involutions* $F \subset B$ *linked by the relation*

$$\bar{\pi}(\tau\omega) = F\,\bar{\pi}(\omega)\,F\ , \tag{B,14}$$

this bijection being given as follows : one has, on the one hand

$$\begin{cases} F = \pi\ (f) \\ \bar{\pi}\ (\omega) = \pi\ (\omega \oplus 0\ f)\quad ,\ \omega \in Q_t\Omega \end{cases} \tag{B,15}$$

and, conversely

$$\pi\,(\omega_o \oplus \omega_1\, f) = \bar{\pi}\,(\omega_0) + \bar{\pi}\,(\omega_1\, F\quad ,\ \omega_0, \omega_1 \in Q_t\,\Omega. \tag{B,16}$$

APPENDIX C. ODD (RESP. EVEN) $t^2/2$-PARATRACES OF Ω AS ODD TRACES OF $Q_t\Omega$ (RESP. $E_t\Omega$)

The notions of parabrackets and paratraces, introduced by Connes in [1], can be defined as (presumably important) items in the general theory of $\mathbb{Z}/2$-graded differential algebras. We coined the name parabracket to stress analogies with the bracket of Lie superalgebras, as shown in the following Lemma [C.2].

[C.1]. Definitions. *Let* $\Omega = (\Omega^+ \oplus \Omega^-, d)$ *be a* $\mathbb{Z}/2$-graded *differential* k-algebra, k $= \mathbb{R}$ *of* $\mathbb{C}$, *with* $\partial\omega$ *the grade of* $\omega \in \Omega^+ \cup \Omega^-$.

(i) The **t-parabracket** $\{\ \}_t$, $t \in k$, *is the bilinear product on* Ω *defined as follows*

(C.1) $\{\omega,\omega'\}_t = [\omega,\omega'] - t(-1)^{\partial\omega} d\omega d\omega', \quad \begin{cases} \omega \in \Omega^+ \cup \Omega^- \\ \omega' \in \Omega \end{cases}$

ii) The **t-paratraces** *of* Ω *are the linear forms of* Ω *which vanish on all t-parabrackets* .

[C.2]. Lemma. *With the definitions and notations of [C.1] we have the following rules* :

(C.2) $\{\omega',\omega\}_t = (-1)^{\partial\omega\partial\omega'} \{\omega,\omega'\}_t - t(-1)^{\partial\omega'} \{d\omega,d\omega'\}_t$, $\omega, \omega' \in \Omega^+ \cup \Omega^-$

and

(C.3) $$\begin{aligned}\{\omega,\psi_1\psi_2\}_t = \{\omega\psi_1,\psi_2\}_t & \\ + (-1)^{(\partial\omega+\partial\psi_1)\partial\psi_2}\{\psi_2\omega,\psi_1\}_t & \qquad , \omega, \psi_1, \psi_2 \in \Omega^+ \cup \Omega^- \\ + (-1)^{\partial\psi_1}\{\omega d\psi_1, d\psi_2\}_t & \\ - t(-1)^{\partial\omega}\{d\omega d\psi_1, \psi_2\}_t. &\end{aligned}$$

The proof is a straightforward verification. Property (C,2) shows that a linear form ϕ of Ω is a paratrace if it vanishes on either $\{\omega,\omega'\}_t$ or $\{\omega',\omega\}_t$ for each unoriented pair $\omega,\omega' \in \Omega$.

Property (C,3) allows to construct recursive proofs for differential algebras Ω generated in low dimension.

[C.3] Proposition . *Let* Ω *be a* Z/2- graded *differential algebra with grading involution* θ *and differential* d. *And let* $Q_t\,\Omega$ *(with product* $\overset{*}{t}$ *and involution* τ, cf.(B,1),(B,3)) *and* $\mathcal{E}_t\Omega$ *be defined as in* **[B.1]** , *resp.* **[B.2]** *of Appendix B. Let, in addition* $\theta^{\pm}$ *and* $\tau^{\pm}$ *be the projections on the* $\substack{even\\odd}$ *parts of* Ω, *resp* . $Q_t\,\Omega$:

(C.4)
$$\begin{cases} \theta_{\pm} = \dfrac{1 \pm \theta}{2} \\ \tau_{\pm} = \dfrac{1 \pm \tau}{2} = \theta_{\pm} \pm \dfrac{t}{2}\theta\ d \end{cases},$$

so that

(C.5)
$$\theta_{\pm}\,\tau_{\pm} = \theta_{\pm}\,(1 + \frac{t}{2}\,d) = \theta_{\pm} + t\frac{d}{2}\,\theta_{\mp}$$

Then

(i) *There is a bijection between the odd* [46] $t^2/2$- paratraces ω *of* Ω *and the odd* [47] *traces* T *of* $Q_t\,\Omega$, *given as follows* : ω *is obtained from* T *as*

(C.6)
$$\nu(\omega) = \begin{cases} 0\,, & \omega \in \Omega^{+} \\ T(\omega)\,, & \omega \in \Omega^{-} \end{cases},$$

and (as a consequence) T *from* ω *as*

(C.7)
$$T = \nu \circ \theta^{-}\,\tau^{-} = \nu \circ \tau^{-}$$

i.e.

(C.7a)
$$T(\omega) = \begin{cases} \dfrac{t}{2}\,\nu\,(d\omega)\,, & \omega \in \Omega^{+} \\ \nu(\omega)\,, & \omega \in \Omega^{-} \end{cases}$$

---

[46] for the grading θ of Ω ; i.e. vanishing on Ω^{+} , or s.t. $\nu \circ \theta^{-} = \nu$

[47] for the grading τ of $Q_t\Omega$;i.e. such that $T \circ \tau = -T$, or $T = T \circ \tau^{-}$

(ii) *there is a bijection between the even* [48] $t/2$- paratraces ν *of* Ω *and the odd* [49] *traces* $\mathbb{T}$ *of* $E_t\Omega$ *such that* $\mathbb{T} = \mathbb{T} \circ \tau$, *given as follows* : ν *is obtained form* $\mathbb{T}$ *as*

$$\text{(C.8)} \quad \nu(\omega) = \begin{cases} \mathbb{T}\,(\,(0,\omega)\,), & \omega \in \Omega^+ \\ 0 & , \omega \in \Omega^- \end{cases}$$

and $\mathbb{T}$ *from* ν *as*

$$\text{(C.9)} \quad \begin{cases} \mathbb{T}\,(\,(\,.\,,\,o\,)\,) = 0 \\ \mathbb{T}\,(\,(\,o,\,.\,)\,) = \nu \circ \theta^+\,\tau^+ \end{cases}$$

i.e.

$$\text{(C.9a)} \qquad \mathbb{T}((\omega_0,\omega_1)) = \begin{cases} \nu\,(\omega_1)\,,\ \omega_1 \in \Omega^+ \\ \\ \frac{t}{2}\nu\,(d\omega_1), \omega_1 \in \Omega^- \end{cases} , \omega_0 \in \Omega\,.$$

(in fact (C.8) *yields a linear map from the odd traces* $\mathbb{T}$ *of* $\mathcal{E}_t\Omega$ *to the even* $t^2/2$-paratraces ν *of* Ω, *the restriction of this map to the traces of* $\mathcal{E}_t\Omega$ *such that* $\mathbb{T} = \mathbb{T} \circ \tau$ *being injective and onto).*

For the proofs, we refer to [6], Appendix C

[C.4] Remark. *The even and odd parts of parabrackets are sums of parabrackets, according to the rules*

$$\text{(C,10)} \qquad \theta_\pm\,\{\omega,\omega'\}_t = \{\theta_+\omega,\,\theta_\pm\omega'\}_t + \{\theta^-\omega,\,\theta_\pm\omega'\}_t,\ \omega,\omega' \in \Omega$$

Accordingly, the even and odd parts of any paratrace is itself a paratrace. These result hold in fact if one replaces θ *by any grading of the algebra* Ω *commuting or anticommuting with the differential* d. The proof is a straightforward verification (cf. Remark **[D.4]** below).

---

[48] for the grading θ of Ω ; i.e. vanishing on Ω^- , or s.t. $\nu \circ \theta^+ = \nu$

[49] for the grading $\hat{\tau}$ of $\mathcal{E}_t\Omega$; i.e. vanishing on all $(\omega,o) \in \mathcal{E}_t\Omega$, $\omega \in \Omega$.

APPENDIX D. THE BASIC MAPS RELATING Ω AND $Q_t\Omega$

In what follows Ω,d, θ, $Q_t\,\Omega$,and τ are as in **[B.1]**

[D.1] **Lemma** (i) *The projections* $\theta_\pm$ *and* $\tau_\pm$ *on* $\Omega^\pm$, *resp.* $Q_t\Omega^\pm$ (cf. (C,1)) *are given by* [50]

(D,1)
$$\begin{cases} \theta_\pm = \tau_\pm \mp \frac{t}{2}\tau d = \tau_\pm \pm \frac{t}{2} d\tau \\ \tau_\pm = \theta_\pm \pm \frac{t}{2}\theta d = \theta_\pm \mp \frac{t}{2} d\theta \end{cases}$$

they fulfill

(D,2)
$$\begin{cases} d\theta_\pm = \theta_\mp\, d \\ d\tau_\pm = \tau_\mp\, d \end{cases}$$

and combine to yield the operators

(D,3)
$$I_\pm = \tau_\pm\, \theta_\pm = (1 - \tfrac{t}{2} d)\, \theta_\pm = \tau_\pm\, (1 - \tfrac{t}{2}\, d)$$

(D,4)
$$J_\pm = \theta_\pm\, \tau_\pm = \theta_\pm (1 + \tfrac{t}{2} d) = (1 + \tfrac{t}{2} d)\, \tau_\pm$$

(D,5)
$$K_\pm = \theta_\pm - I_\pm = \tau_\mp\, \theta_\pm = -\,\theta_\mp\, \tau_\pm = J_\pm - \tau_\pm$$
$$= \tfrac{t}{2}\, \theta_\mp\, d = \tfrac{t}{2} d\theta_\pm = \tfrac{t}{2} d\tau_\pm = \tfrac{t}{2} \tau_\mp\, d$$

(ii) $I_\pm$ *and* $J_\pm$ *are idempotents with the following kernels and images*

(D,6)
$$I_\pm^2 = I_\pm \quad , \text{Ker}\, I_\pm = \Omega^\mp \quad , \ \text{Im}\, I_\pm = Q_t\, \Omega^\pm$$

(D,7)
$$J_\pm^2 = J_\pm \quad , \text{Ker}\, J_\pm = Q_t\, \Omega^\mp \quad , \ \text{Im}\, J_\pm = \Omega^\pm$$

---

[50] We formulated throughout two statements which correspond to the respective choices of the upper, resp. lower sign $\pm$, $\mp$ in the formulae below.

moreover they fulfill

(D,8) $$J_\pm I_\pm = \theta_\pm \tau_\pm \theta_\pm = \theta_\pm \quad , J_\pm I_\mp = 0$$

(D,9) $$I_\pm J_\pm = \tau_\pm \theta_\pm \tau_\pm = \tau_\pm \quad , I_\pm J_\mp = 0$$

Consequently we have

(D,10) $$\Omega = Q_t\Omega = \Omega^+ \oplus \Omega^- = Q_t\Omega^+ \oplus Q_t\Omega^- = \Omega^+ + Q_t\Omega^- = \Omega^- \oplus Q_t\Omega^+$$

and $I_\pm|_{\Omega^\pm}$, *resp.* $J_\pm|_{Q_t\Omega^\pm}$, *are inverse bijections relating* $\Omega^\pm$ *and* $Q_t\Omega^\pm$.

(iii) θ *and* τ *coincide on the kernel of* d, *which is a graded subspace of* Ω (*and* $Q_t\Omega$) *on which* $I^\pm$ *and* $J^\pm$ *reduce to the identity. For* $t \neq 0$ *the* $\substack{even \\ odd}$ *part of* Ker d *equals*

(D,11) $$(\text{Ker } d)^\pm = \Omega^\pm \cap Q_t\Omega^\pm$$

(iv) $K_\pm$ *is nilpotent, with, for* $t \neq 0$, *the following kernel and image*

(D,12) $$K_\pm^2 = K_\pm K_\mp = 0, \text{ Ker } K_\pm = \Omega^\mp \oplus (\text{Ker } d)^\pm, \text{ Im } K_\pm = d(\Omega^\mp) = (d\Omega)^\pm$$
$$= Q_t\Omega^\mp \oplus (\text{Ker } d)^\pm$$

(v) *We have*

(D,13) $$\begin{cases} K_\pm = K_\pm I_\pm = K_\pm J_\pm = I_\mp K_\pm = J_\mp K_\pm = - I_\mp I_\pm = J_\mp J_\pm \\ I_\pm K_\pm = J_\pm K_\pm = K_\pm I_\mp = K_\pm J_\mp = 0 \end{cases}$$

The above situation in the case $t \neq 0$ is subsumed in the figure

(D,14)

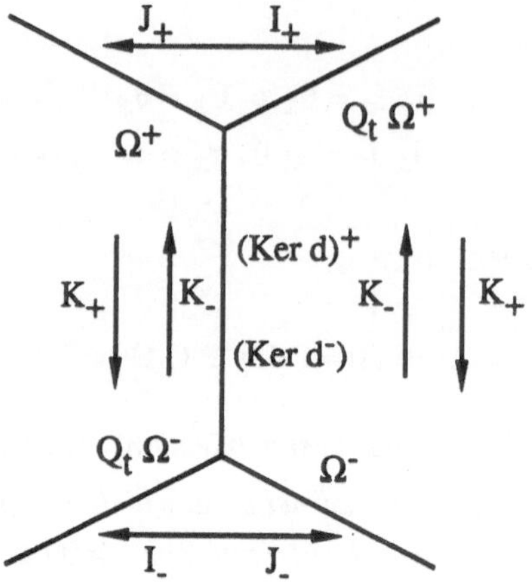

where

(D,14) $\begin{cases} \Omega^+ = \quad , Q_t\Omega^+ = \quad , \text{ with intersection } (\text{Ker d})^+ = \\ Q_t\Omega^- = \quad , \Omega^- = \quad , \text{ with intersection } (\text{Ker d})^- = \end{cases}$

Proof : (i) (D,1), (D,2) follow from (B,3a) and (C,4) ; and imply (D,3) through (D,5).

(ii) (D,8), (D,9) follow from (D,3), (D,4) using $1 - \frac{t}{2}d = (1 + \frac{t}{2}d)^{-1}$ Ker $I_\pm \subseteq$ Ker $\theta_\pm$ (resp. Ker $J_\pm \subseteq$ Ker $\tau_\pm$) follow from (D,3) and (D,6) (resp. from (D,4) and (D,7)). (D,8), resp. (D,9) then implies that $I_\pm$ (resp. $J_\pm$) are idempotents with the stated images.

(iii) $\tau = \theta(1 + td)$ coincides with θ on Ker d, which is invariant under d owing to $d\theta = -\theta d$. Let $\omega \in \Omega^\pm \cap Q_t\,\Omega^\pm$ then $\tau\omega = (1 - td)\,\theta\omega = \pm\,\omega = \pm\,(1 - td)\omega$, hence $d\omega = 0$, proving the inclusion $\supset$ in (D,11). The inverse inclusion was evident from the fact that $\theta = \tau$ on Ker d.

(iv) Nilpotency of $K_\pm$ follows from $d^2 = 0$. The statement about Im $K_\pm$ is obvious from (D,5), which on the other hand implies, for $\omega \in \Omega$:

(D,15) $$\omega \in \text{Ker } K_\pm \Leftrightarrow d\theta_\pm\,\omega = 0 \Leftrightarrow \theta_\pm\,\omega \in (\text{Ker d})^\pm$$
$$\Leftrightarrow d\tau_\pm\,\omega = 0 \Leftrightarrow \tau_\pm\,\omega \in (\text{Ker d})^\pm$$

whence the stated expressions for Ker $K_\pm$.

(v) straight forward from (D,3) through (D,5).
The noted properties of $I_\pm$ and $J_\pm$ now imply

[D.2] ***Corollary*** *: Let* $\Omega^{*\pm} \cong \Omega^{\pm *}$ *be the set of linear forms on* Ω *vanishing on* $\Omega^{\mp}$:

(D,16) $$\Omega^{*\pm} = \{\phi \in \Omega^* \, ; \, \phi = \phi \circ \theta^{\pm}\}$$

and let $(Q_t\Omega)^{*\pm} \cong (Q_t\Omega^{\pm})^*$ *be the set of linear forms on* $Q_t\Omega$ *vanishing on* $Q_t\Omega^{\mp}$:

(D,17) $$(Q_t\Omega)^{*\pm} = \{\bar\phi \in Q_t\Omega^* \, ; \, \bar\phi = \bar\phi \circ \tau^{\pm}\} \ ,$$

one has the following inverse linear bijections between those sets :

(D,18) $$\begin{cases} \phi = \bar\phi \circ I_\pm \\ \bar\phi = \phi \circ J_\pm \end{cases} \ , \quad \begin{cases} \phi \in \Omega^{*\pm} \\ \bar\phi \in Q_t\Omega^{*\pm} \end{cases} \ ,$$

corresponding to the commutative diagram

(D,19) $$\begin{array}{ccc} \Omega^{\pm} & \overset{I_\pm}{\underset{J_\pm}{\rightleftarrows}} & Q_t\,\Omega^{\pm} \\ \phi & & \bar\phi \\ & \mathbb{C} & \end{array}$$

These bijections can also be characterized by requiring

(D,20) $$\begin{cases} \phi(\omega) = \begin{cases} \bar\phi\,(\omega) & , \ \omega \in \Omega^{\pm} \\ 0 & , \ \omega \in \Omega^{\mp} \end{cases} \\ \\ \bar\phi\,(\omega) = \begin{cases} \phi\,(\omega) & , \ \omega \in Q_t\Omega^{\pm} \\ 0 & , \ \omega \in Q_t\Omega^{\mp} \end{cases} \end{cases} \ , \quad \begin{cases} \phi \in \Omega^{*\pm} \\ \bar\phi \in Q_t\Omega^{*\pm} \end{cases}$$

or

$$(D,20b)\qquad \begin{cases} \phi(\omega) = \begin{cases} \bar{\phi}(\omega) & , \omega \in Q_t\Omega^{\pm} \\ -\frac{t}{2}\bar{\phi}(d\omega), & \omega \in Q_t\Omega^{\mp} \end{cases} \\ \bar{\phi}(\omega) = \begin{cases} \phi(\omega) & , \omega \in \Omega^{\pm} \\ \frac{t}{2}(d\omega), & \omega \in \Omega^{\mp} \end{cases} \end{cases} , \quad \begin{cases} \phi \in \Omega^{*\pm} \\ \bar{\phi} \in Q_t\Omega^{*\pm} \end{cases}$$

this agreeing with the following characterizations of $\Omega^{*\pm}$ *and* $Q_t\Omega^{*\pm}$ *:*

$$(D,21)\qquad \phi \in \Omega^{*\pm} \Leftrightarrow \begin{cases} \phi(d\omega) = 0 & , \omega \in Q_t\Omega^{\pm} \\ \phi(\omega) = -\frac{t}{2}\phi(d\omega) & , \omega \in Q_t\Omega^{\mp} \end{cases}$$

$$(D,22)\qquad \bar{\phi} \in Q_t\Omega^{*\pm} \Leftrightarrow \begin{cases} \bar{\phi}(d\omega) = 0 & , \omega \in \Omega^{\pm} \\ \bar{\phi}(\omega) = \frac{t}{2}\phi(d\omega) & , \omega \in \Omega^{\mp} \end{cases}$$

Proof : The inverse bijections : ϕ , $\bar{\phi}$ in (D,18) are are the transpose of the inverse bijections $I_{\pm}$, $J_{\pm} : \Omega^{\pm} \leftrightarrow Q_t\,\Omega^{\pm}$ (cf. (D,19)). The relations (D,20) immediately follow from $I_{\pm} = \tau_{\pm}\,\theta_{\pm}$, $J_{\pm} = \theta_{\pm}\,\tau_{\pm}$. The characterizations (D,21), (D,22) of $\Omega^{*\pm}$ immediately follow from (D,1).

The next Lemma shows how the maps $I_{\pm}$ and $I_{\pm}$ turn $t\,^2/2$ - parabrackets of Ω^-, resp. Ω^+, into commutators of $Q^t\Omega^-$, resp. $E_t\Omega^-$, which by transposition yields the results of Appendix C.

[D.3] ***Lemma.*** *We have the following identities : for* $t \in \mathbb{C}$.

$$(D,23)\qquad I_-\{\omega,\omega'\}_{t^2/2} = [I_+\omega,^* I_-\omega']_- + [I_-\omega,^* I_+\omega']_- - [K_+\omega,,^* K_-\omega']_- + [K_-\omega,^*_, K_+\omega']_- \quad ,\omega,\omega' \in \Omega$$

$$(D,24)\qquad J_-[\psi^*_;\psi']_- = \{J_+\psi, J_-\psi'\}_{t^2/2} + \{J_-\psi, J_+\psi'\}_{t^2/2} + \{K_+\psi, K_-\psi'\}_{t^2/2} - \{K_-\psi, K_+\psi'\}_{t^2/2} \quad ,\psi,\psi' \in Q_t\Omega$$

(D,25) $$I_+\{\omega,\omega'\}_{t^2/2} = [\![I_+\omega ,^* I_+\omega']\!]_- + [\![I_-\omega ,^* I_-\omega']\!]_-$$
$$- [\![K_+\omega \,{}^*_{;}\, K_+\omega']\!]_- + [\![K_-\omega ,^* K_-\omega']\!]_-$$
$$(= [I_+\omega ,^* I_+\omega']_- - [K_+\omega ,^* K_+\omega']_-$$
$$+ [I_-\omega ,^* I_-\omega']_- + [K_-\omega \,{}^*_{;}\, K_-\omega']_-)$$

(D,26) $$J_+ [\![\psi \,{}^*_{;}\, \psi']\!]_- = \{J_+\psi, J_+\psi'\}_{t^2/2} + \{J_-\psi, J_-\psi'\}_{t^2/2}$$
$$+ \{K_+\psi, K_+\psi'\}_{t^2/2} - \{K_-\psi, K_-\psi'\}_{t^2/2}$$

using the notation

(D,27) $$[\psi \,{}^*_{;}\, \psi']_\pm = \psi \,{}^*_{t}\, \psi' \pm \psi' \,{}^*_{t}\, \psi$$

$$, \psi, \psi' \in Q_t\Omega$$

(D,28) $$[\![\psi \,{}^*_{;}\, \psi']\!]_\pm = \psi \,{}^*_{t}\, \psi' \pm \psi' \,{}^*_{t}\, \tau\psi$$

For the proof, we shall use the obvious

[D.4]*Remark.* *We have the rules*

(D,29) $$\begin{cases} \theta^+(\omega\omega') = (\theta^+\omega)(\theta^+\omega') + (\theta^-\omega)(\theta^-\omega') \\ & , \omega, \omega' \in \Omega \\ \theta^-(\omega\omega') = (\theta^+\omega).(\theta^-\omega') + (\theta^-\omega).(\theta^+\omega') \end{cases}$$

(D,30) $$\begin{cases} \tau^+(\psi \,{}^*_{t}\, \psi') = (\tau^+\psi) \,{}^*_{t}\, (\tau^+\psi') + (\tau^-\psi) \,{}^*_{t}\, (\tau^-\psi') \\ & , \psi, \psi' \in Q_t\Omega \\ \tau^-(\psi \,{}^*_{t}\, \psi') = (\tau^+\psi). {}^*_{t} (\tau^-\psi') + (\tau^-\psi). {}^*_{t} (\tau^-\psi') \end{cases}$$

and hence also the variants of there rules where the product in (D,29) *is replaced by the parabracket* $\{\}_{t^2/2}$ *(cf.(C,1) ; and where the product* $\underset{t}{*}$ *in (D,30) is replaced by the corresponding commutator* (D,27)[51] .

Check of (D,23) : we have, for $\omega, \omega' \in \Omega$, using [D,4]

(D,31) $$\theta_-\{\omega,\omega'\}_{t^2/2} = \{\theta_+\omega,\theta_-\omega'\}_{t^2/2} + \{\theta_-\omega,\theta_+\omega'\}_{t^2/2}$$

$$= (\theta_+\omega)(\theta_-\omega')-(\theta_-\omega')(\theta_+\omega) - \frac{t^2}{2}(d\theta_+\omega)\,(d\theta_-\omega')$$

$$+ (\theta_-\omega)(\theta_+\omega')-(\theta_+\omega')(\theta_-\omega) + \frac{t^2}{2}(d\theta_-\omega)\,(d\theta_+\omega')$$

$$= [\theta_+\omega \underset{t}{*} \theta_-\omega'] + t(\theta_-\omega')(d\theta_+\omega) - \frac{t^2}{2}(d\theta_+\omega)\,(d\theta_-\omega')$$

$$+ [\theta_-\omega \underset{t}{*} \theta_+\omega']_- - t\,(\theta_-\omega')(d\theta_+\omega) + \frac{t^2}{2}(d\theta_-\omega)\,(d\theta_+\omega')$$

$$= [\theta_+\omega \underset{t}{*} \theta_-\omega']_- - \frac{t^2}{2}\,[d\theta_+\omega \underset{t}{*} d\theta_-\omega')] + t\,(\tau_-\theta_-\omega')\,(d\theta_+\omega)$$

$$+ [\theta_-\omega \underset{t}{*} \theta_+\omega']_- - t\,(\tau_-\theta_-\omega)(d\theta_+\omega)$$

$$= [\theta_+\omega \underset{t}{*} \theta_-\omega']_- - 2\,[K_+\omega^*, K_-\omega') + 2\,(\tau_-\theta_-\omega')\,(\tau\theta_+\omega)$$

$$+ [\theta_-\omega \underset{t}{*} \theta_+\omega'] - 2\,(\tau_-\theta\_\omega)(\tau\_\theta_+\omega')$$

hence applying τ_+ on both sides

51 or, for that matter, the modified product (D,28)

(D,32) $$I_-\{\omega,\omega'\}_{t^2/2} = [I_+\omega \circledast I_-\omega']_- + [K_+\omega \circledast K_-\omega']$$

$$-2[K_+\omega \circledast K_-\omega']_- + [K_-\omega \circledast K_+\omega'] + [I_-\omega \circledast I_+\omega']$$

Check of (D,24) : we apply J_- to both sides of (D,23) where we insert $\omega = J_+\psi + J_-\psi$, $\omega' = J_+\psi' + J_-\psi'$, $\psi, \psi' \in Q_t\Omega$.
We then have that $I_\pm\omega = \tau_\pm\psi$, $\theta_\pm\omega = J_\pm\psi$, and analogously for ω' and ψ', hence, using **[D.4]** and (,)

(D,33) $$J_- I_- \{\omega,\omega'\}_{t^2/2} = \theta_-\{\omega,\omega'\}_{t^2/2} = \{\theta_+\omega,\theta_-\omega'\}_{t^2/2} + \{\theta_-\omega,\theta_+\omega'\}_{t^2/2}$$

$$= \{J_+\psi, J_-\psi'\}_{t^2/2} + \{J_-\psi, J_+\psi'\}_{t^2/2}$$

(D,34) $$J_-\{[I_+\omega \circledast I_-\omega']_- + [I_-\omega \circledast I_+\omega']_-\}$$

$$= J_-\{[\tau_+\psi \circledast \tau_-\psi]_- + [\tau_-\psi \circledast \tau_+\psi']_-\} = J_-\tau_-[\psi \circledast \psi']_-$$

$$= J_-[\psi \circledast \psi']_-$$

(D,35) $$J_-[K_+\omega \circledast K_-\omega']_- = \theta_-\tau_-[K_+\omega \circledast K_-\omega']_-$$

$$= \theta_-[K_+\omega \circledast K_-\omega']_- = \theta\{K_+\omega, K^-\omega'\}_{t^2/2}$$

$$= \{K_+\omega, K_-\omega'\}_{t^2/2} = \{K_+J_+\psi, K_-J_-\psi'\}_{t^2/2}$$

$$= \{K_+\psi, K_-\psi'\}_{t^2/2}$$

(D,36) $$J_-\left[K_-\omega \mathbin{\ast\!\!;} K_+\omega'\right]_- = -J_-\left[K_+\omega' \mathbin{\ast\!\!;} K_-\omega\right] = -\left\{K_+\psi', K_-\psi\right\}_{t^2/2}$$

$$= \left\{K_-\psi, K_+\psi'\right\}_{t^2/2}$$

thus (D,23) yields

(D,37) $$\left\{J_+\psi, J_-\psi'\right\}_{t^2/2} + \left\{J_-\psi, J_+\psi'\right\}_{t^2/2}$$
$$= J_-[[\psi \mathbin{!\!;} \psi']_- - \left\{K_+\psi, K_-\psi'\right\}_{t^2/2} = \left\{K_-\psi, K_+\psi'\right\}_{t^2/2}$$

we proved (D,24).

Check of (D,25) : we have, for $\omega,\omega' \in \Omega$, using [D.4]

(D,38) $$\theta_+\{\omega,\omega'\}_{t^2/2} = \{\theta_+\omega, \theta_+\omega'\}_{t^2/2} + \left\{\theta_-\omega, \theta_-\omega'\right\}_{t^2/2}$$

$$= \left[\theta_+\omega \mathbin{\ast\!\!;} \theta_+\omega'\right]_- - \frac{t^2}{2}\left(d\theta_-\omega\right)\left(d\theta_+\omega'\right)$$

$$+ \left[\theta_-\omega \mathbin{\ast\!\!;} \theta_-\omega'\right]_- - t\left(\theta_-\omega\right)\left(d\theta_-\omega'\right)$$

$$- t\left(d\theta_-\omega'\right)\left(d\theta_-\omega\right) + \frac{t^2}{2}\left(\theta_-\omega\right)\left(d\theta_-\omega'\right)$$

hence applying τ_+ on both sides

(D,39) $I_+ \{\omega,\omega'\}_{t^2/2} = [I_+\omega \overset{*}{t} I_-\omega']_+ + [K_+ \omega \overset{*}{t} K_+\omega']_- - 2\tau_+\{(\tau_-\theta_+\omega) \overset{*}{t}(\tau_-\theta_+\omega')\}$

$+[I_-\omega \overset{*}{t} I_-\omega']_+ + [K_-\omega \overset{*}{t} K_-\omega']_+ -2\tau_+\{(\theta_-\omega) \overset{*}{t}(\tau_+\theta_-\omega')\}$

$-2\tau_+\{(\theta_-\omega) \overset{*}{t}(\tau_+\theta_-\omega')\} +2\tau_+\{(\tau_+\theta_-\omega)(\tau_+\theta_-\omega')\}$

$= [I_+\omega \overset{*}{t} I_+\omega']_- + [K_+\omega \overset{*}{t} K_+\omega']_- 2(K_+\omega) \overset{*}{t} (K_+\omega')$

$+ [I_-\omega \overset{*}{t} I_-\omega']_+ + [K_-\omega \overset{*}{t} K_-\omega']_+ - 2(K_-\omega) \overset{*}{t} (K_-\omega')$

$-2(K_-\omega') \overset{*}{t} (K_-\omega) + 2(K_-\omega) \overset{*}{t} (K_-\omega')$

$= [I_+\omega , I_+\omega']_- - [K_+\omega \overset{*}{t} K_+\omega']_+$

$+ [I_-\omega \overset{*}{t} I_-\omega']_- - [K_-\omega \overset{*}{t} K_-\omega']_-$

Check of (D,26) : we apply J_+ to both sides of (D,25) where we insert as above $\omega = J_+\psi+J_-\psi, \omega'=J_+\psi'+J_-\psi'$, $\psi,\psi\in Q_t\Omega$. We have

(D,40) $J_+I_+\{\omega,\omega'\}_{t^2/2} = \theta_+\{\omega,\omega'\}_{t^2/2} = \{\theta_+\omega,\theta_+\omega'\}_{t^2/2} + \{\theta_-\omega,\theta_-\omega'\}_{t^2/2}$

$= \{J_+\psi,J_+\psi'\}_{t^2/2} + \{J_-\psi,J_-\psi'\}_{t^2/2}$

(D,41) $J_+\{[I_+\omega \overset{*}{t} I_-\omega']_- + [I_-\omega \overset{*}{t} I_-\omega']_-\}_{t^2/2}$

$= J_+\{[\tau_+\psi \overset{*}{t} \tau_+\psi']_- + [\tau_-\psi \overset{*}{t} \tau_+\psi']_-\}_{t^2/2}$

$= J_+\tau_+[\psi \overset{*}{t} \psi']_- = J_+ [\psi \overset{*}{t} \psi']_-$

(D,42) $$J_+[K_\pm\omega \,\ast_{\!,}\, K_\pm\omega']_- = \theta_+\tau_+ [K_\pm\omega \,\ast_{\!,}\, K_\pm\omega']_-$$

$$= \theta_+ [K_\pm\omega \,\ast_{\!,}\, K_\pm\omega']_- = \theta_+ [K_\pm\omega \,\ast_{\!,}\, K_\pm\omega]_\pm$$

$$= \theta_+ [K_\pm\omega \,,\, K_\pm\omega']_\pm = [K_\pm\omega \,,\, K_\pm\omega]_\pm = [K_\pm\omega, K_\pm\omega']$$

$$= \{K_\pm\omega, K_\pm\omega'\}_{t^2/2} = \{K_\pm\psi, K_\pm\psi'\}_{t^2/2}$$

thus (D,25) yields

(D,43) $$\{J_+\psi, J_+\psi'\}_{t^2/2} + \{J_-\psi, J_-\psi'\}_{t^2/2}$$

$$= J_+ [\psi \,\ast_{\!,}\, \psi']_- - \{K_+\psi, K_+\psi'\}_{t^2/2} + \{K_-\psi, K_-\psi'\}t^2/2)$$

APPENDIX E. AN ALTERNATIVE FORM OF THE CHARACTER OF A θ- SUMMABLE MODULE
(Added in july 1988)

This appendix describes a form of the Chern character due to Jaffe, Leniewski and Osterwalder [21].

[E.1] Let $(\mathcal{H},D)$ be a θ-summable Fredholm A-module now taken in the graded sense, i.e. $a \to (a)$ is a $*$-representation of the $\mathbb{Z}/2$-graded $*$-algebra[52] $A = A^0 \oplus A^1$ by bounded operators on $\mathcal{H} = \mathcal{H}^0 \oplus \mathcal{H}^1$ which is grade-preserving :

$$\begin{cases} \varepsilon(a)\varepsilon = (a) & , a \in A^0 \\ \varepsilon(a)\varepsilon = -(a) & , a \in A^1 \end{cases} \tag{E,1}$$

(in other terms $a \to (a)$ is a zero grade $*$-homomorphism : $A \to \mathcal{B}(\mathcal{H})$ of graded algebras, where $\mathcal{B}(\mathcal{H})$ is graded by the involution $\theta = \mathrm{Ad}\varepsilon$). We set $D^2 = H$, so that e^{-Ht} is trace class for all $t > 0$.

For $B \in \mathcal{B}(\mathcal{H})$ s.t. $B\mathcal{D}_D \subset \mathcal{D}_D$ we set

$$\delta B = i\,[D,B] = i\varepsilon\,[\varepsilon F,B]_- \tag{E,2}$$

with [,] a graded, and $[\ ,\ ]_-$ an ordinary commutator. The **supertrace** Str is defined as follows : for B trace class, we set

$$\mathrm{Str}(B) = \mathrm{Tr}\{\varepsilon B\} \qquad , B \in BL^1(\mathcal{H}) \tag{E,3}$$

δB is then a graded derivation, and Str a graded trace (i.e. Str vanishes on graded commutators)

We note the properties

[52] A is assumed unital with unit 1, with (1) = 1.

(E,4) $$\mathrm{Str} \circ \theta = \mathrm{Str}$$

hence, for B_1, B_2 homogeneous elements

(E,5) $$\mathrm{Str}(B_1, B_2) = (-1)^{\partial B_1 \partial B_2}\, \mathrm{Str}\,(B_2, B_1)$$

$$= (-1)^{\partial B_1}\, \mathrm{Str}\,(B_2, B_1) = (-1)^{\partial B_2}\, \mathrm{Str}\,(B_2, B_1)$$

further

(E,6) $$\mathrm{Str} \circ \delta = 0$$

and[53]

(E,7) $$\delta^2 = -\mathrm{Ad}H$$

i.e.

(E,8) $$\delta^2 B(t) = \dot{B}(t) \quad , t \in \mathbb{R}$$

where

(E,9) $$B(t) = e^{-Ht}\, B\, e^{Ht}$$

We note that

(E,10) $$\|B\|_* = \|B\| + \|\delta B\|$$

is an algebraic $*$-norm on the domain of δ. We set, moreover, for $B_1, ..., B_n \in \mathcal{B}(\mathcal{H})$, $n \in \mathbb{N}$, and $t \in \mathbb{R}$[54]

---

[53] equality of maps with coïncidence of domains.

[54] $\overset{t}{f}(n)$ is defined generally by (E,11) as a functional of n maps $B_i : t \in \mathbb{R} \to B_i(t) \in \mathcal{B}(\mathcal{H})$, $i = 1, ..., n$; and in particular, for the funcitons (E,9) where $B_i \in \mathcal{B}(\mathcal{H})$, $i = 1, ..., n$.

(E,11)
$$\begin{cases} f^t_{(n)}(B_1,\ldots,B_n) = \displaystyle\int_{0\le t_1\le \ldots \le t_n \le t} B_1(t_1)\ldots B_n(t_n)dt \\ \qquad = \displaystyle\int_0^t dt_n \int_0^{t_n} dt_{n-1}\ldots\int_0^{t_2} dt_1 B_1(t_1)\ldots B_n(t_n) \\ f^t_{(0)} = 1 \end{cases}$$

We then have that

[E.2] ***Proposition.*** *Setting for* $a_0, a_1, \ldots, a_n \in A, t \in \mathbb{R}$:

(E,12)
$$\tilde{\phi}^t_n(a_0 da_1 \ldots da_n) = t^{-\frac{n}{2}} \operatorname{Str}\left\{ (a_0) f^t_{(n)}(\delta(a_1), \ldots, \delta(a_n)) e^{-tH} \right\}$$

defines a normalized entire cyclic cocycle of A, in the sense that we have

(E,13)
$$\tilde{\phi} \circ \mathbb{B}\, \lambda = \tilde{\phi} \circ \mathbb{B} = \tilde{\phi} \circ \beta\, \varepsilon \quad (\text{i.e.} \lambda\, B_0 \tilde{\phi} = B_0 \tilde{\phi} = b\tilde{\phi})$$

The proof will follow from a sequence of Lemmas

[E.3] Lemma *We have that, with* $1<k<n$; $n = 1, 2, \ldots$.

(E,14)
$$\begin{cases} f^t_{(n)}(B_1,\ldots,B_{n+1},\hat{B}_n) = f^t_{(n-1)}(B_1,\ldots,B_{n-1})B_n(t) \\ \quad - f^t_{(n-1)}(B_1,\ldots,B_{n-2},B_{n-1}B_n) \\ f^t_{(n)}(B_1,\ldots,\hat{B}_k,\ldots,B_n) = f^t_{(n-1)}(B_1,\ldots,B_kB_{k+1},\ldots,B_n) \\ \quad - f^t_{(n-1)}(B_1,\ldots,B_{k-1}B_k,\ldots B_n) \\ f^t_{(n)}(\hat{B}_1,B_2,\ldots,B_n) = f^t_{(n-1)}(B_1B_2,B_3,\ldots,B_n) \\ \quad - B_1(0)\, f^t_{(n-1)}(B_2,\ldots,B_n) \end{cases}$$

Proof : straightforward verification from (E,11)

[E.4] Lemma. *Let for* $B_0, \ldots, B_n \in \mathcal{B}(\mathcal{H})$, $t \in \mathbb{R}$:

$$\text{(E,15)} \quad F^t_{(n)}(B_0, B_1, \ldots, B_n) = \mathrm{Str}\left\{ B_0 f^t_{(n)}(B_1, \ldots, B_n) e^{-tH} \right\}$$
$$= \int_{s_0+s_1+\ldots+s_n=t} \mathrm{Str}\,(B_0 e^{-s_0H} B_1 e^{-s_1H} \ldots B_n e^{-s_nH}) ds$$

We have that

$$\text{(E,16)} \qquad F^t_{(n)}(1, B_1, \ldots, B_n) = t\, F^t_{(n-1)}(B_1, \ldots, B_n)$$

and, for $1 \leq k < n$

$$\text{(E,17)} \begin{cases} F^t_{(n)}(B_0, B_1, \ldots, \hat{B}_n) = F^t_{(n-1)}(\theta(B_n) B_0, B_1, \ldots, B_{n-1}) \\ \qquad - F^t_{(n-1)}(B_1, \ldots, B_{n-2}, B_{n-1} B_n) \\ F^t_{(n)}(B_0, \ldots, \hat{B}_k, \ldots, B_n) = F^t_{(n-1)}(B_0, \ldots, B_k B_{k+1}, \ldots, B_n) \\ \qquad - F^t_{(n-1)}(B_0, \ldots, B_{k-1} B_k, \ldots B_n) \end{cases}$$

further for B_i *of grade* ∂B_i, $i = 1, \ldots, n$

$$\text{(E,18)} \qquad F^t_{(n)}(B_0, B_1, \ldots, B_n) = 0 \text{ unless } \sum_{k=0}^{n} \partial B_k = 0$$

$$\text{(E,19)} \qquad F^t_{(n)}(B_n, B_0, \ldots, B_{n-1}) = (-1)^{\partial B_n} F^t_{(n)}(B_0, B_1, \ldots, B_n)$$

Proof : (E,16) and (E,19) follow from the second equation (E,15) immediate consequence of the first. (E,17) is a rewriting of (E,14), whist (E,18) follows from (E,4).

[E.5] Lemma. *We have that, for* $a_0, a_1, \ldots, a_n \in A^0 \cup A^1$, $y \in \mathbb{R}$:

$$(E,20)\quad (-1)^{\partial a_0} (a_0)\, \delta f^{\,t}_{(n)}(\delta(a_1), \ldots, \delta(a_n))e^{-tH}$$

$$= -\delta\left\{ (a_0) f^{\,t}_{(n)}(\delta(a_1), \ldots, \delta(a_n))e^{-tH} \right\} + \delta(a_0) f^{\,t}_{(n)}(\delta(a_1), \ldots, \delta(a_n))e^{-tH}$$

$$= (-1)^{\partial a_0} (a_0 a_1)\, f^{\,t}_{(n-1)}\left(\delta(a_2), \ldots, \delta(a_i a_{i+1}), \ldots \delta(a_n)\right)e^{-tH}$$

$$+ \sum_{i=1}^{n-1} (-1)^{i + \sum_{k=0}^{i} \partial a_k} (a_0)\, f^{\,t}_{(n-1)}\, (\delta(a_1), \ldots, \delta(a_i a_{i+1}), \ldots \delta(a_n))e^{-tH}$$

$$-(-1)^{(1+\partial a_n)\left(n-1+\sum_{k=0}^{n-1} \partial a_k\right)} (a_n a_0)\; f^{\,t}_{(n-1)}\left(\delta(a_1), \ldots, \delta(a_{n-1})\right)e^{-tH}$$

$$+(-1)^{n+\sum_{k=0}^{n-1} \partial a_k} \left[(a_0)\; f^{\,t}_{(n-1)} \left(\delta(a_1), \ldots, \delta(a_{n-1})\right)e^{-tH}, (a_n)\right]$$

hence upon taking the supertrace

$$(E,21)\qquad (b\tilde{\phi}^{\,t}_{n-1})\, (a_0 da_1, \ldots, da_n) = t^{-\frac{n-1}{2}}\; F^{\,t}_{(n-1)}\, (\delta(a_0), \delta(a_1), \ldots \delta(a_n))$$

Proof : The first equality (E,20) is the derivation property of δ applied to a product of the three factors (a_0), $f^{\,t}_{(n)}\,(\delta(a_1), \ldots, \delta(a_n))$ and e^{-tH}, the latter vanishing under δ. The last equality then arises by computing the first term r.h.s., using again the derivation property of δ[55], this yielding n terms the ith of which obtained by replacing $\delta(a_i)$ by $\delta^2(a_i) = (a_i)$[56] : we then apply the rules (E,14), whence the stated sum of terms in the second equation (E,20).

---

[55] cf.(E,11)

[56] cf. (E,8)

Upon taking the supertrace, the first term second line[57] and the last term in (E,20) vanishes : the sum in the second equation (E,20)then yields the l.h.s. of (E,21), since one has, owing to (1,16)

(E,22)
$$\beta(a_0 da_1 \dots da_n = (-1)^{\partial a_0} a_0 a_1 \; da_2 \dots da_n$$

$$+ \sum_{i+1}^{n-1} (-1)^{i + \sum_{k=0}^{i} \partial a_k} a_0 da_1 \dots d(a_i a_{i+1}) \dots da_n$$

$$- (-1)^{(1+\partial a_n)(n-1+\sum_{k=0}^{n}\partial a_k)} a_n a_0 da_1 \dots da_{n-1}$$

[E.6] ***Lemma.*** *We have that*

(E,23)
$$\left(B_0 \tilde{\phi}^{\,t}_{n-1}\right)(a_0 da_1, \dots, da_n) = t^{-\frac{n-1}{2}} F^{\,t}_{(n)}(\delta(a_0), \delta(a_1), \dots \delta(a_n))$$

$$= \left(\lambda B_0 \tilde{\phi}^{\,t}_{n-1}\right)(a_0 da_1, \dots, da_n)$$

Proof: We have, from (1,30a)

(E,24)
$$\mathbb{B}_0 (a_0 da_1, \dots, da_n) = 1\, da_0 \dots da_n + (-1)^{n + \sum_{k=0}^{i} \partial a_k} a_0 da_1, \dots, da_n d1$$

hence (cf. (E,12)) since $\delta 1 = 0$, and using (E,16)

(E,25)
$$(\tilde{\phi}^{\,t}_{n+1} \circ \mathbb{B}_0)(a_0 d_1, \dots, da_n) = t^{-\frac{n+1}{2}} F^{\,t}_{(n)}(1, \delta(a_0), \delta(a_1), \dots \delta(a_n))$$

$$= t^{-\frac{n-1}{2}} F^{\,t}_{(n)}(\delta(a_0), \delta(a_1), \dots \delta(a_n))$$

---

[57] cf.(E,6)

Furthermore we have, from (1,21), and

$$\text{(E,26)}\quad \lambda(a_0 da_1, \dots da_n) = (-1)^{(1+da_n)\left(n+\sum\limits_{k=0}^{i} \partial a_k\right)} a_n\, da_0 da_1 \dots da_{n-1} \ ,$$

hence, taking account of (E,18) and (E,19), since $(1+\partial a_n)\partial a_n = 0 \bmod 2$

$$\text{(E,27)}\quad (\tilde{\phi}_{n-1} \circ \mathbb{B}_0 \lambda)(a_0 da_1, \dots, da_n) = t^{-\frac{n-1}{2}} (-1)^{(1+\partial a_n)} F^{\,t}_{(n)} (\delta(a_n), \delta(a_0), \dots, \delta(a_{n-1}))$$

$$= t^{\frac{n-1}{2}} F^{\,t}_{(n)} (\delta(a_0), \delta(a_1), \dots, \delta(a_n))$$

[E.7] Remarks

(i) *The character* $\tilde{\phi}^t = (\tilde{\phi}^{\,t}_{(n)})$ *stems as indicated in* (3,3) *from a normalized entire cyclic cocycle* ϕ^t *satisfying*

$$\text{(E,28)}\qquad \begin{cases} \phi^t \circ \mathbb{B}_0 \Lambda^1 = o \\ (b+B)\phi^t = o \end{cases}$$

(ii) *By restriction to even (resp. odd) N-parity and even (resp. odd) intrinsic parity the character* ϕ^t *splits in four cocycles of the types* (6,9) *and* (6,10) .

(iii) *Regauging* ϕ_n *as follows :*

$$\text{(E,29)}\qquad \underline{\phi}_n (a_0, a_1, \dots, a_n) = (-1)^{\sum\limits_{k\,odd} \partial a_k + n \sum\limits_{k=0}^{n} \partial a_k}\, \phi_n (a_0 da_1, \dots da_n)$$

one get a hierarchy of multilinear forms on A fulfilling the handier cocycle condition

$$\text{(E,30)}\qquad (\underline{b} + \underline{B})\, \underline{\phi} = 0$$

where[58]

$$(E,40)\quad (\underline{b}\phi)^{n+1}(a_0,a_1,\ldots,a_{n+1}) = \sum_{i=0}^{n}(-1)^i \phi_n(a_0,\ldots,a_ia_{i+1},\ldots,a_{n+1})$$

$$-(-1)^{(1+\partial a_{n+1})(n+\sum_{k=0}^{n}\partial a_k)}\phi_n(a_na_0,a_1,\ldots,a_n)$$

and $\underline{B} = \underline{\lambda}\, B_0$ *where*

$$(E,41)\qquad (\underline{\lambda}\,\phi)_n(a_0,a_1,\ldots,a_n) = (-1)^{(1+\partial a_n+1)(n+\sum_{k=0}^{n}\partial a_k)}\phi_n(a_n,a_0,\ldots,a_{n-1})$$

[E.8] Remark. The 1-paratrace μ^t associated to the normalized cyclic cocycle ϕ^t as in (3,3) is as follows : one has, for $\tilde{a}_0 \in \tilde{A} = A \oplus \mathbb{C}\tilde{1}$, $a^1, \ldots, a_n \in A$:

$$(E,8)\quad \mu_n^t(\tilde{a}_0 da_1 \ldots sa_n) = t^{-\frac{n}{2}}\,\mathrm{Str}\left\{(\tilde{a}_0)\, f_{(n)}^t(\delta(a_1)\ldots\delta(a_n))e^{-tH}\right\}$$

In fact the above Lemmas [E.5] and [E.6] yield in combination with Remark [4.4] a direct proof of the paratrace property of μ^t.

---

58 $\underline{b}$ and $\underline{\lambda}$ are the same as b resp. λ in (A,5) resp. foonote 37. (see [7] Remark [3.5] and Ap. K or 12

APPENDIX F. THE CHARACTER OF GRADED KMS - FUNCTIONALS (Added in august 1988)

The Jaffe et al. character described in Appendix F can be generalized in two ways [22]. First, the "super-Boltzman" functional can in fact be replaced by any "super KMS" functional w.r.t. to a "supersymmetric" modular group, this revealing a basic relationship between cyclic cohomology and the KMS structure. Second, this can be done irrespective of parity.

[F.1] Definition. Let $A = A^\circ \oplus A^1$ be a $\mathbb{Z}/2$graded Banach algebra.

(i) a one-parameter automorphism group $t \to \alpha_t$ of A is **supersymmetric** whenever (a) it commutes with the grading, (b) it has an infinitesimal generator "with a square root" in that it is the square of an odd derivation δ of A[59])

(ii) with (α,δ) as in (i), a linear functional φ of A is **graded** (or **super**)-t-KMS, $t \in \mathbb{R}$, whenever

(F,1) $\varphi(ba) = (-1)^{\partial a \partial b}\, \varphi\,\{a\, \alpha_{it}(b)\}$

for all α-differentiable $a,b \in A^\circ \cup A^1$.

We then have

[F,2] Theorem. *Let* t *be a super* t-KMS *functional of* $A = A^\circ \oplus A^1$, $t \in \mathbb{R}$, *and set* [60] *for* $a_0, a_1, \ldots, a_n \in A$

(F,2) $\phi^t(a_0 da_1 \ldots da_n) = t^{\frac{n}{2}} i^n \displaystyle\int_{0 \le t_1 \le \ldots \le t_n \le t} \varphi\,\{a_0\, \alpha_{it_1}(\delta a_1) \ldots \alpha_{it_n}(\delta a_n)\}dt$

then ϕ^t *is a (non normalized) cyclic cocycle in the sense* [F,13] [61] .

[F.3] Remark. *In a field theory situation , we know from* [23] *that each extremal* t - KMS *state* φ *of* A° *extends to a weakly clustering state of* A *fulfilling*

(F,3) $\varphi(ba) = \varphi\,\{a(\alpha_{it} \circ \gamma)(b)\}$

with γ *the identity or the grading automorphism of* A. *But* (F,3) *is a parent of* (F,1). Does this really depend upon an assymptotically invariant assumption ?

---

[59] with due consideraton of domains

[60] definition in cyclic cohomology within the differential envelope. For the definition in terms of multilinear forms cf. [F.].

[61] hence an entire cyclic cocycle for a φ with adequate growing conditions

Bibliography

[0] A. Connes. Non-Commutative Differential Geometry. I. The Chern character in K-homology II. De Rham Cohomology and Non-commutative Algebra. Pub. Math. IHES 237(1985).

[1] A. Connes. Entire cyclic cohomology of Banach algebras and characters of θ-summable Fredholm modules. IHES preprint M/87/46 (Nov. 1987). K-Theory I, 519 (1988).

[2] A. Connes. Compact metric spaces, Fredholm modules and hyperfinitness. Pacific J. Math. 137, 225 (1989).

[3] A. Connes. The Dixmier trace and the Wodcicki residue. IHES preprint. see also the action functional in non-commutative geometry C.M.P. 17, 673 (1988).

[4] see J. Bellissard's report : C*-algebras in solid state physics, Marseille, preprint CPT 87/P.2047.

[5] A. Jaffe, A. Lesniewski and J. Weitsman. Index of a family of Dirac operators on loop space. Comm. Math. Phys. **112**, 75 (1987).

[6] D. Kastler. Entire cyclic cohomology of Z/2-graded Banach algebras. CPT -88/P.2087Marseille Preprint. (Jan. 1988).

[7] D. Kastler. Cyclic cohomology within the differential envelope. An introduction to Alain Connes' non commutative differential geometry. Travaux en cours; Ed. Scient. Herman, Paris (1987).

[7a] D. Kastler. Z/2-graded cyclic cohomology withing the differential envelope. Proceeding of the June 1985 Iowa conference. Contemporary mathematics vol. 62 (1987).

[7b] D. Kastler. Introduction to Alain Connes non-commutative differential geometry. XXIInd Karpacz Winter School. Fields and Geometry (A. Jadczyk ed.) World Scientific (1986).

[8] D. Kastler. An invitation to Alain Connes' cyclic cohomology Trends and developments in the eighties. Bielefeld Encounters IV and V (S. Albeverio and Ph. Blanchard eds.). World Scientific (1985).

[9] J. Cuntz. A new look at KK-theory. K-theory I 31-51 (1987).

[10] R. Zekri. A new description of Kasparov's theory of C*-algebra extensions. Journ. Funct. Anal. 84, 441 (1989).

[11] A. Connes and J. Cuntz. Quasi homomorphismes, cohomologie cyclique et positivité. C.M.P. 114, 515 (1988).

[12] R. Coquereaux and D. Kastler. Remarks on the differential envelopes of differential algebras. Pacific J. Math. 137, 245 (1989).

[13] J. Loday and D. Quillen. Cyclic homology and Lie algebra homology of matrices. Comm. Math. Helv. **59**. (1984).

[14] A. Connes Cohomologie cyclique et foncteurs Ext^n. CR. Acad. Sci. Paris **296** 953 (1983).

[15] B.L. Tsigan. Homology of matrix algebras and the Hoschild homology. Usp. Math. Nauk **38** 217 (1983).

[16] P. Seibt. Cyclic homology of algebras, World Scientific (1987).

[17] D. Burghelea. Cyclic homology and the algebraic K-theory of spaces I. Contemporary Mathematics **55** Part I (1986) II. Topology **25** 303 (1986).

[18] C. Kassel Cyclic homology, comodules and mixed complexes in print J. Algebra.

[19] J.D.S. Jones Cyclic homology and equivariant homology Invent. Math. **87** 403 (1987).

[20] R. Bott and L. Tu. Differential forms in algebraic topology . Graded texts in Mathematics 82, Springer Verlag.

[21] A. Jaffe, A. Lesniewski, K. Osterwalder, Quantum K. Theory. I. The Chern character Harvard preprint HUTMP B209 (March 1988).

[22] D. Kastler. Cyclic cocycles from (graded) KMS-functionals.CPT-88/P.2148 August 1988.

[23] H. Araki, R. Haag, D. Kastler, M. Takesaki. Extension of KMS States and Chemical Potential. Comm. Math. Phys. 53, 97 (1977).

STOCHASTIC ANALYSIS, MATHEMATICAL STRUCTURES AND DISTRIBUTIONS METHODS

Paul KREE
UA 213 CNRS et
Departement de Mathématiques
Université de Paris 6
Place Jussieu, 75005 Paris

à Raphael HOEGH KROHN, à Michel SIRUGUE

" Il y a un nombre indéfini (et peut être même infini) de volumes dans la bibliothèque de Babel. Ces ouvrages sont écrits en diverses langues mais avec les mêmes signes à savoir le point, la virgule, l'espace et les lettres de l'alphabet... On ne peut former aucune suite de caractères par exemple

d h c m r ℓ c h t d j

que la divine bibliothèque n'ait déjà prévue, et qui dans quelqu'une de ses langues secrètes ne renferme une signification terrible ... Finalement la bibliothèque est désertée par ses lecteurs bien que éclairée, solitaire, infinie, parfaitement immobile, armée de volumes précieux, inutile, incorruptible, secrète..."

J.L. BORGES, La Bibliothèque de Babel

The author thanks J. Dieudonné (resp. E. Carlen) for fruitful discussions concerning structures (resp. stochastic analysis) ; and also Ph. Blanchard and Testard for invitation at this very interesting Marseille conference. In 1933, A.N. Kolmogoroff did found probability theory as geometry and as algebra as a mathematical discipline on the structure of abstract probability space $(\Omega, \mathcal{Q}, P)$.

This has been a great discovery. Nevertheless one generally believes today that this foundation presents some incompatibility with physics (see 1). Another argument is that today's stochastic analysis constructed on these foundations appears as the union of an increasing number of non interacting theories spraying out in all directions constructed in very different ways ; this without substantial hope concerning physics. The goal of this

S. Albeverio et al. (eds.), Stochastics, Algebra and Analysis in Classical and Quantum Dynamics, 153–168.

lecture is first to show that what is really in question is not Kolmogoroff's foundations but how these foundations have been understood and applied since fifty years, and also the fact that integration theory is not consistently and effectively structured at the present time. Guided by Nelson's idea and methods of structures and distributions, we propose a solution to this second problem. Finally using this optic, we will propose elsewhere [KRE 90] a foundation of stochastic analysis in the line of [KR 89]; showing that this analysis is not really diverging but constructing progressively a stochastic calculus extending and improving the classical one and also working in physics.

1. THE PROBLEMS OF FOUNDATION OF PROBABILITY THEORY AND OF STRUCTURATION INTEGRATION THEORY

1.1 Probability theory

1.1.1 We start with the following two questions : Does the probabilistic phenomena treated in today's physical science, appear as relevant to a mathematical probability theory ? Can a probability theory effectively be presented today in a way privileging systematically set theoretical integration theory ? (as frequently done in today's probabilistic literature).

1.1.2 D. Hilbert did consider mechanics and probability theory as two connected parts of physics where mathematics plays a preponderant role since he formulated in 1900 his sixth problem in a context of statistical mechanics and in the following terms :
"Les recherches sur la géométrie nous conduisent au problème : traiter sur ce modèle les branches de la physique où les mathématiques jouent un rôle prépondérant : avant tout le calcul des probabilités et la mécanique".
Notice incidentally that the history of science does not tell us if this problem has been solved concerning mechanics, and in this case who did solve it. In any case the revolution in physics generated by Einstein's idea immediately showed, soon after 1900 that the main mathematical problem connected with mechanics was really not to formulate axioms for classical mechanics organized in the following way

$$\text{Newton mechanics} \rightarrow \text{Lagrangian mechanics} \rightarrow \text{Hamilton mechanics} \qquad (1.1)$$

and axioms for classical fields ... The main mathematical problem was to develop new mathematical tools suggested by relativity theory and to develop relativistic dynamics.
Note also that the mechanics literature would be unreadable if written in terms of axioms and as a part of mathematics because in physics mathematics only gives mathematical models of the real world. Hence physical principles explaining how physical concepts are

modelized (i.e. represented in models) and also explaining the physical meaning of theorems are needed. Note also that the development of a physical theory naturally leads to the construction of successive mathematical models built up introducing progressively new mathematical tools. For example the three modellings (I.1) corresponds to the introduction in mechanics of the following tools

differential equations → calculus of variations → symplectic forms.

One sees in this way the difficulty of axiomatic approaches in the Hilbert spirit, since it is not helping synthetic views, since axioms introduce an artificial rigidity, without putting necessarily in light the main physical concepts. In other words what is needed for the mathematical foundation of a physical theory is not only good basic structures, but also good physical principles (making short synthesis of the good ideas we learn from physicists), and freedom for the introduction of new models and corresponding mathematical structures.

1.1.3 In 1979, J. Dieudonné tells us that considering the general architecture of today's mathematics, he was convinced that there was some difficulties connected with the formulation of Kolmogoroff's axioms and that probability theory was not built up in an optimal way. J. Dieudonné also explained to us that Bourbaki's optic in mathematics is not exactly in the line of Hilbert's idea. Let us underline this important fact. D. Hilbert did suggest practically independent structurations of different parts of mathematics and also of some parts of physics, suggesting in this way independent developments of these parts. But N. Bourbaki did realize in 1940 - 1960 an interacting and consistent structuration of fundamental mathematics, known in 1930 - 1950, which is optimal since coding the scientific information with the good signs and in the right way since suggesting interaction of structures and making in evidence in each particular case the right methodology. This did generate an interacting and consistent development of mathematics and its applications. A first example is Euclidean geometry ; the famous "A bas Euclide !" means that the approach of geometry using vector spaces is more effective than an approach of Euclidean type since the power of all mathematical structures connected with vector spaces is at your disposal. But this sentence also means that Einstein's revolutionary views in physics (hence in particular in geometry) would be impossible if geometry would be uniquely modelized in Euclidean terms : and in fact the power of structures connected with vector spaces have been used by Einstein in order to modelize mechanics, relativistic classical fields... A second example in mechanics where the following problems appeared during the first half of this century.

(i) discover the foundamental geometric structures of mechanics like manifolds (modelling the configuration space) symplectic manifolds (modelling phase spaces) Lie groups (modelling symmetries of the configuration space)...

(ii) Using these models and mechanical ideas, find how these fundamental geometric structures interact. For example Noether's theorem holds if some symmetry group preserves the Lagrangian.

(iii) Find an efficient presentation of the theory of the fundamental mechanical ideas and also of the mathematical results, but also allowing the free interaction, with all mathematical structures because the real world is so complex that structures cannot be fixed a priori.

These structuration problems of mechanics have been solved in the sixties ; this produced the foundation of today's mechanics : see the works of A. Lichnerovicz and Palais, Abraham's book (Foundations of mechanics 1968) J.M. Souriau's book (Structure des systèmes dynamiques. Dunod 1970...). In books of mechanics written later in the same spirit, the synthesis between physics and mechanics is so perfect that adding more physical comments you obtain books for practicians, and that bypassing these comments, you have the impression to read an excellent book of differential geometry. In other words an important part of mathematics (differential geometry) is progressing using the same structures and close ideas. The next example shows inconvenients of independent structurations of different parts of physics.

1.1.4 The point of view of quantum physicists. The following three parts of physics have been structured independently and in fact using very different mathematical structures :

a) Quantum Mechanics by Von Neumann in 1932. The structure of Hilbert space and some structures connected with unbounded operators of Hilbert spaces are fundamental in Von Neumann's theory. In this theory the physical concept of probability and Euclidean - Galilean modeling of space-time appears as secondary, since only used for the final formulation of results.

b) Probability theory by A.N. Kolmogoroff in 1933. The Grundbegriffe [KOL 33] suggest that for each situation involving random phenomena, the corresponding modelling necessarily involves fundamentally some abstract probability space $((\Omega,\mathcal{A},P)$, such that the family of events of interest can be canonically identified with some family $\mathcal{A}$ of subsets of Ω. Also [KOL 33] may suggest that set theoretical integration theory must be the fundamental mathematical method for the study of this model.

c) the theory of relativistic quantum fields created by Wightman and Gårding in 1956. Quantum phenomena are fundamentally treated here as in a), but c) presents two main differences with respect to a). First since a relativistic quantum field is a relativistic concept, space and time are modelized as in special relativity theory. Second since relativistic quantum fields are not defined punctually, these fields are modelized by vectorial distributions.

These axiomatics habe been undoubtly great discoveries. But since the Hilbertian modelling of quantum phenomena used in a) (hence also in c)) only involves the quantum kinematics, one generally believes today that a more realistic modelling involving quantum dynamics can be done using functional integration : see R.P. Feynman and Schwinger for the physical approach and mainly E. Nelson for the existing mathematical approach, mixing Kolmogoroff's methods with other ones : in particular mathematical analysis, since the Ω of interest are not arbitrary sets but locally convex Hausdorff spaces having some nice topology. Hence a basic contradiction since mathematicians have constructed three very different theories for the study of very close physical phenomena. Note that there is presently less hope for a consistent approach of relativistic quantum phenomena on the line of c) since there are fundamental contradictions between today's concept of relativistic quantum field with Einstein's relativity theory. But, the non relativistic approach of quantum phenomena is considered as physically consistent : hence corresponding quantum dynamics is in course of elaboration on the basis of Nelson's idea (works of E. Nelson, P. Blanchard, P. Combe , W. Zheng, E. Carlen, J.C. Zambrini). But the existing stochastic analysis is even insufficient for the elaboration of this non relativistic theory.

1.1.5. E. Nelson writes in the introduction of n°117 of Annals of Math. Studies (Princeton 1987) :

"I am sure that many probabilists teaching a beginning graduate course, have also the feeling that these measure theoretic foundations (of probability theory) serve more to salve our mathematical consciences that to provide an incisive tool for the scientist who wishes to apply probability theory".

1.1.6. M.M. Rao analyses carefully in [RAO 88] paradoxes arising in Kolmogoroff's probability theory, since the conditioning by events with vanishing probability is not allowed there. This is forbidden even if such conditionings arise naturally in physics, in mechanics...

1.1.7. In general engineers physicists... studying random phenomena are confronted with these paradoxes. They have other difficulties since probability theory gives not all the results they need. In general they read with difficulties the probabilistic litterature because no physical comment which may be used as landmark is given there.

1.1.8. In conclusion the answers to our two questions are negative. Probabilistic concepts fundamentally involve physics, in very various ways. The present situation is very confusing since the expression "probability theory" denotes today two very different things : physical probability theory viewed as a theory of phenomena involving probability ; and mathematical probability theory as existing today i.e. many times isolated from physics and other mathematical theories. Also many physicists and analysts (in particular the author of this lecture) have been perhaps misled by the systematic formulation of mathematical probability theory in set theoretical terms since this formulation is hiding and weakening the fantastic power of probabilistic concepts in physics and in mathematics especially as mixed with other mathematical structures. This fact is clearly appearent in all B. Simon's works developping a new kind of probability theory effective in quantum physics. Also in the introduction of [SIM 74] B. Simon underlines in this respect the exceptional merit of E. Nelson : not only did he recognizes the first this power, be he created a more effective probability theory, with a new style using sharp functional analysis.

1.2 Relation with the structuration of integration theory

Since the general organization of N. Bourbaki's Elements de Mathématiques has been established in 1940-1955, the fact that probability theory has not been considered fundamental in this organization is not surprising since martingales theory, probabilistic potential theory, Ito's stochastic calculus were just being born at that time and since Nelson's work on functional integration began in 1965. Giving numerous specific examples, N. Bourbaki explains in the introduction of his integration theory (Hermann 1952) that his approach is mainly motivated by finite dimensional analysis. Hence this approach only works in finite dimensional analysis even if [BOU 58] gaves soon later fundamental results concerning all topological spaces of interest in dimension free analysis. Since this analysis exists today and since further progresses of this analysis are needed in physics, the elaboration of an unified and dimension free integration theory is a first urgency. This elaboration is studied below.

2 A SHORT REVIEW OF THE BEGINNING OF MODERN INTEGRATION AND PROBABILITY THEORIES

As usual BM(d) denotes the d-dimensional Brownian Motion.

2.1 Integration Theory and Probability Theory before Wiener's Theory of B.M.

1905 Lebesgue's integration theory

1920 Thanks to further works of Radon, Daniell, Frechet ... the basic set theoretical integration theory is well established. Numerous applications on finite dimensional spaces are known but no applications involving infinite dimensional vector spaces are known in physics concerning the applications to probabilities :

1909 Borel and Cantelli introduce for the first time countably additive probabilities.

2.1.1 Before Wiener's theory, the fact that one can associate to any random vector $v = (v_1,\dots,v_d)$ of $\mathbb{R}^d$, a probability measure m on the Borel σ-field $\sigma(\mathbb{R}^d)$ of $\mathbb{R}^d$ is "known" i.e. well understood at the physical level. The measure m usually called the law of $\underline{v}$ or the distribution of $\underline{v}$ can be denoted Law ($\underline{v}$) and called the chance measure of $\underline{v}$ since $\forall A \in \sigma(\mathbb{R}^d)$, m(A) gives a measure of the chance that repeated and independent observations of v give a result ∈ A. It was also known" than any Borel mapping $\ell : \mathbb{R}^d \to \mathbb{R}^e$ (for example ℓ linear) modelizes an observation of $\underline{v}$, and that the random vector $\ell(\underline{v})$ of $\mathbb{R}^e$ has for chance measure the direct image $\ell(m)$ of m by ℓ.

2.1.2 Let x_1, x_2 and x_3 orthonormal coordinates in our Euclidean space. We refer to E. Nelson [NEL 67] for an historical and scientific presentation of b = B.M.(1) i.e. for the Ox_1 - displacement for $t \geq 0$ of a Brownian particle located at the origin at time $t = 0$. We only try to explain how a physicist or a mathematician could understood this phenomenon in the period 1900 - 1923 i.e. without using Kolmogoroff's probability theory (1933). In 1900 Jean Perrin did precise experiments on B.M.. He noticed that the observed trajectories strongly suggest the continuous functions without derivatives constructed by mathematicians. Hence the observed paths of b belong to the space $\Omega_b = C(\mathbb{R}_+)_0$ of all continuous functions f(t) vanishing at time $t = 0$. Also b defines a collection $\{b(t), t \geq 0\}$ of random numbers. Hence for any time intervall $I_j = (t_j, t_{j+1})$, the corresponding Ox_1 - displacement of our particle is the random number

$$\Delta_j b = b(t_{j+1}) - b(t_j) \tag{2.1}$$

This displacement is informally obtained integrating on I_j the function $s \to b'(s)$ giving the Ox_1 - component at any time s of the velocity of our particle. Hence informally

$$\Delta_j b = \int_{t_j}^{t_{j+1}} b'(s)\, ds \tag{2.2}$$

Since the variations of the Ox_1 - component of the velocity of our particle is generated by collisions with water molecules, the function $s \to b'(s)$ cannot be continuous.

In view of Newton's fundamental law of mechanics, since the B.M. is generated by the collisions of our particles with water molecules, the computation of Law $(\Delta_j b)$ needs statistical informations concerning the Ox_1 - component of the Ox_1 - velocities of water molecules at a given point. This is a problem of statistical mechanics tractable using the works of Gibbs, Maxwell and Boltzmann (using analytic methods of mechanics, an ergodic principle...). The final result of Einstein's theory coïncides with the results observed by Jean Perrin ; this can be summarized in the following way.

2.1.3 Physical description of b = B.M.(1)

A1. This is a random continuous function i.e. successive and independent experiments can be repeated on b ; and the observed paths always belong to Ω_b.

A2. Denoting b(t) the random position at any time t we have of course b(0) = 0. Moreover for any increasing and finite family $\dot{t}$ of

$$t_1 = 0 < t_2 < \ldots < t_N \tag{2.3}$$

the corresponding increments $\Delta_j b$ $(1 \leq j < N)$ are independent random numbers such that $\Delta_j b$ is Gaussian centered with variance $t_{j+1} - t_j$.

Remark : The chance measure of the random vector $(b(0), \Delta_1 b, \ldots, \Delta_{N-1} b) \in \mathbb{R}^N$ is the product m' of δ_0 and of the N – 1 Gauss measures

$$\text{Law } (\Delta_j b) = \frac{1}{\sqrt{2\pi\, \delta_j}} \left(\exp - \frac{t^2}{2\delta_j} \right) dt \tag{2.4}$$

with $\delta_j = t_{j+1} - t_j$. Since

$$b(t_j) = \Delta . b + \ldots + \Delta_{j-1}\, b$$

the chance measure $m_{\dot{t}}$ of the random vector

$$b_{\dot{t}} = (b(t_1), \ldots, b(t_N)) \tag{2.5}$$

is the direct image of m' by the following linear mapping of $\mathbb{R}^n$

$$(x_j) \to (y_j) \qquad \text{with} \qquad y_j = x_1 + \ldots + x_j$$

2.2 N. Wiener's physical and mathematical work (1923) on the Brownian Motion

Even if N. Wiener was only working with $\{b(t), t \in [0,1]\}$ we keep the presentation on Ω_b working with $\{b(t), t \geq 0\}$.

2.2.1 N. Wiener first extends 2.1.d for infinite dimensional vector spaces. More precisely he introduces the σ-field $\sigma(\Omega_b)$ on Ω_b generated by all mapping

$$\Omega_b \ni w \xrightarrow{\alpha_{\dot{t}}} (w(t_1) \ldots w(t_n)) = w(\dot{t}) \in \mathbb{R}^N \tag{2.6}$$

N. Wiener first interpret A1 as meaning that it must be possible to associate to b some chance measure P_b on $\sigma(\Omega_b)$. Then he interpret A2 as meaning that $\forall\, \dot{t} = (t_1, \ldots, t_N)$,

$\forall$ A $\in \sigma(\mathbb{R}^N)$ the measure of the chance that $\alpha_{\dot{t}}^*(w) \in A$ is the number $m_{\dot{t}}^*(A)$. In order workds one must have

$$\forall \dot{t} \quad \forall A \qquad P_b\left(\alpha_{\dot{t}}^*(A)\right) = m_{t'}(A) \tag{2.7}$$

Then N. Wiener succeeds to prove that such measure P_b exists !! Wiener's proof appears today as very intricated since not only using the physical considerations sketched before, but also set theoretical measure theory (1905 - 1920) but also (this is less known) some results of Gateaux's and Levy's potential theory on infinite dimensional spaces (1913-1922).

2.2.2 Let us list some mathematical structures arising more or less implicitely in the proof of Wiener's revolutionary result, and also let us indicate the relation of these structures with physics.
Giving two real vector spaces X and U let us denote by X ... U the data of these spaces and also of a duality on X × U we denote x.u = <x,u> = u(x) ... Since for any t > 0, the valuation mapping α_t is defined by the measure δ_t on $\mathbb{R}^+$, the following is true :

2.2.3 <u>Wiener's modelling of b = BM</u>(1) :

N. Wiener represents mathematically b by the following mathematical structure

$$(U_b \dots \Omega_b ; P_b) \tag{2.8}$$

where U_b = Span (δ_t , t > 0) denotes the subspace of Ω_b generated by the valuation mappings δ_t.
More precisely he modelizes the random function b by the triple (Ω_b , $\sigma(\Omega_b)$, P_b) and he modelizes the family {b(t) , t ≥ 0} of random numbers by the results of the observation of b defined by the linear forms δ_t.
The derivative b' of b is implicit in [WIE 23] and is the main physical motivation of two Wiener's later works, for instance Wiener's stochastic integration defined by Wiener as the linear isometry

$$H_{b'} \ni \mathbb{1}_{(t_j,t_{j+1})} \rightarrow \Delta_j b \in L^2(\Omega_b) \tag{2.9}$$

where $H_{b'} = L^2(\mathbb{R}_+,dt)$, and he denotes this isometry

$$\varphi \rightarrow \int_0^{\infty} \varphi(t)\, db(t) \tag{2.10}$$

Also b' is strongly related with Wiener's decomposition in chaos. But since Schwartz distribution theory did not exist at that time, Wiener was unable to modelize b'. Let us recall how this can be done today, by a structure isomorphic with (2.8). Endowing Ω_b with its natural Frechet topology, the derivation in the sense of distributions in $\mathcal{D}'$ =

$\mathcal{D}(]0,+\infty[)'$ maps Ω_b onto the subspace $\Omega_{b'}$ of $\mathcal{D}'$ consisting of the distributions derivatives w' of all $w \in \Omega_b$. The map $D : \Omega_b \to \Omega_{b'}$ defined in this way is bijective since $w' = 0$ and $w(0) = 0$ imply $w = 0$. Hence carrying over structures by D a Frechet structure and a probability measure $P_{b'}$ on $\Omega_{b'}$ modelling b' viewed as a random distribution. The transpose of D induces a bijection of some subspace $U_{b'}$ of the dual of $\Omega_{b'}$ onto the subspace U_b of the dual of Ω_b. One easily sees

$$D^T(\mathbb{1}_{(0,t)}) = \delta_t \qquad \text{hence} \qquad U_{b'} = \text{Span}\left\{\mathbb{1}_{(0,t)} \; ; \; t > 0\right\}$$

Hence we obtain the diagram

$$\begin{array}{cccc} (\; U_b & \ldots\ldots & \Omega_b & ; \; P_b \;) \\ D^T \uparrow & & \downarrow D & \\ (\; U_{b'} & \ldots\ldots & \Omega_{b'} & ; \; P_{b'} \;) \end{array} \tag{2.11}$$

whose second line gives a modelling of b'. In this modelling the increments $\Delta_j b$ are represented by the linear forms $\mathbb{1}_{(t_j, t_{j+1})}$ on $\Omega_{b'}$.

2.3 Kolmogoroff's Grundbegriffe [KOL 33]

2.3.1 In 1933 Kolmogoroff did found probability theory as a mathematical theory using set theoretical integration theory, and in particular the structure $(\Omega, \mathcal{Q}, P)$ of probability space. The connection with physics is only discussed in two pages in [KOL 33]. But nevertheless this has also been very useful in the standard applications since events, random variables ... can be represented mathematically using Kolmogoroff's theory.

2.3.2 A second important novelty of [KOL 33] is to underline strongly the power of set theoretical integration theory for the treatment of probabilistic problems. For example giving some intervall J, Kolmogoroff defines a stochastic process on J as a family $\{\zeta_t , t \in J\}$ of random variables. Combining this definition with Borel-Cantelli lemma, he did found with a simpler proof a counterpart of Wiener's results.

2.3.3 A third important novelty of [KOL 33] is the theory of conditional expectations.

2.3.4 A necessary counterpart of this strong underlining of the power of $(\Omega, \mathcal{Q}, P)$ has been that other structures have disappear, for example all those explicit or implicit in Wiener's theory and in Daniell's theorem. Also the physical aspects have disappear. For example, Kolmogoroff makes no distinction between the probabilized spaces of interest in physics (alway's having an additional Lusin topological structure) and "abstract probabilized spaces" only arising as technical devices in proofs. Also the same terminology of stochastic process is used for marginal processes of random elements (as $\{b(t), t \geq 0\}$ in Wiener's case) and abstract stochastic processes defined in terms of "abstract probabilized spaces". Hence there was the risk for the newborn theory of a divorce with physics and also with the mathematical structures modelling corresponding physical concepts.

3. A SHORT REVIEW OF TODAY'S INTEGRATION AND PROBABILITY THEORIES

3.1 Introduction

3.1.1 Let us imagine a mathematician or a mathematical physicist beginning today researchs in integration theory and in stochastic analysis. Since the results sketched in 2 have more than fifty years, since structures interact freely in mathematics and since physics suggests the introduction of more and more structures in a given theory, he can believe a priori that integration theory and probability theory have progressed during these fifty years in the following way :

- since todays the distinction between mathematic and mathematical physics is much more clear that at Kolmogoroff time, a formulation of probability theory closer from physics has been adopted
- all additional structures needed for the development of the theory or of applications have been introduced, leading progressively to the study of much more complex structures that $(\Omega,\mathcal{A},m)$
- the mathematical results only involving the interaction with topology or with the usual Newton-Leibnitz differential calculus on $\mathbb{R}^d$ have been discovered very rapidly (for example twenty years).

He will also expect that today's probability theory has been build up and constructed as all mathematical theories in general i.e. according the following formulation rule and problematic.

3.1.2 Formulation Rule (FR) : giving some mathematical result mr, find without any a priori constraint the structures and the corresponding methods allowing the most effective and the most general formulation of mr.

3.1.3 Problems (PB) motivating these mathematical results are suggested by the theory or by the applications.

In fact the situation is very different.

3.1.4 A first reason of this difference of probability theory is the exceptional power of the structure $(\Omega, \mathcal{Q}, m)$; and this has been forecasted by Kolmogoroff. More precisely even if this structure appears as very simple, the corresponding theory is deep, and the interaction with other mathematical theory is very deep. Let us give two examples

a) The interaction with topology gaves almost immediately the Riesz representation theorem for the compact spaces [0,1]. But for more general topological spaces the first results only appear ≈ 1940. Finally this interaction has only been clarified after 1965 : see for ex. [BIL 68], [SCH 73] [FER 67].

b) Semi martingales theory established finally in 1974-1979 is only written with very few signs : those of set theoretical integration theory and Newton - Leibnitz differential calculus. This differential calculus is needed in order to write for any $\mathbb{R}^d$ - valued semi-martingale (ζ_t) and any C^2-function f defined on $\mathbb{R}^d$ the formula of change of variable (also called Ito's formula).

Semi-martingales theory whose first versions have been published in 1974-1980 is the final result and the synthesis of works of numerous and brillant probabilists during fourthy years. The theory began in 1937 with the discovery by P. Levy of the concept of martingale, with Ito's stochastic calculus (1946-1955) with Doob's martingales theory (1950) with Hunt and Doob probabilistic potential theory... Then further improvements and progresses have suggested the introduction of the concept of semi-martingales (Doleans-Dade and Meyer 1970) and showed that all the "stochastic calculus" is in fact semi-martingales theory. Even if this theory can be summarized in few pages this is a fantastically deep, and powerful theory. For example classical potential theory and L^p estimates cannot more be studied today without using the stochastic calculus.

3.1.5 Semi-martingales theory has been studied untill 1980 precisely in the framework of Kolmogoroff's probability theory i.e. applying the following constrained variant of F.R. and

FR' : Giving some probabilistic result pr with statement and proof formulated in terms of set theoretical integration theory, find another probabilistic result still formulated in the same way, with a weaker hypothesis and same kind of proof.

PB' : Problems leading to these results are mainly those suggested by the theory.

Hence an optimal coding of semi-martingales theory for specialists. Since during the period of elaboration of this theory, physics did suggest other problems, another stochastic analysis has been created in view of these problems. Hence two very different kinds of stochastic analysis.

3.2. Classical Stochastic Analysis (Cl. S.A.)

3.2.1 Classical stochastic analysis appears as centered on semi-martingales theory. For the part concerning directly this theory, this is natural. But perhaps in order to stay apparently strictly conform with Kolmogoroff's axioms some interactions of other mathematical theories have been retranscripted completing FR' by the following reformulation rule.

FR" : Giving some physical or mathematical theory Th interacting with semi-martingales theory and producing some result r find the weakest reformulation in set theoretical terms of a part of Th (or even of a part of another existing theory Th') allowing by interaction with semi-martingales theory a proof of r (or of a reformulation of r in set theoretical terms).

For example quantum mechanics can be reformulated in this way but one unfortunately loses physical applications. Also ten calculi at least have been created in this way : for example the calculus taught by Malliavin is so tricky that the genesis and the real nature of this calculus have only been recently understood [STR 88][KRE 89]. In any way an enormous "probabilistic" litterature has been constructed. Therefore this litterature is of very difficult access since presenting an hudge of mathematical informations coded only with very few signs, but using very different (and some times unknown!) coding rules.

3.3. Modern Stochastic analysis (Mod. S.A.)

3.3.1 The corresponding litterature is presented as usually in mathematics i.e. according to the formulation rule FR.
This analysis is more in the line of [KOL 33] + [WIE 23] than in the line of [KOL 33] hence is more close from physics and structures. We make an incomplete review of modern stochastic analysis beginning with its main part.

3.3.2 Mathematical theory of functional integration and of quantum dynamics. These theories are strongly connected with statistical mechanics and also with fundamental problems of today's quantum physics. Unfortunately the stochastic calculus of cl. S.A. is no working more here since the processes (ξ_t; t real) of interest take values in infinite dimensional state spaces (no more in $\mathbb{R}^d$) or take values in non commuting algebra, or involve a process $f(\xi_t)$ where f is no more C^2. Hence completely new idea are needed. The basic physical idea and the basic mathematical idea have been discovered by E. Nelson. These theories are usually presented in a physical context and use much more functional analysis than cl. S.A. : see for example all B. Simon's books.

3.3.3 Interaction of semi-martingales theory with differental geometry.
We refeer for example to D. Elworthy's work and to the vivifying Rogers and William's book.

3.3.4 Quantum probability theory of Accardi, Frigerio, Von Waldenfels developps non commuting versions of cl. S. A. substituing here the classical integration theory by the non commuting version of I. Segal for example : see the seminar published in the Lectures Notes.

NB Perhaps the qualificative "non commutative" would be more appropriate than "quantum" since quantum probability may suggests that usual probability theory cannot be applied to quantum physics and this is not true.

3.3.5 The non commutative stochastic calculus of Hudson and Parthasarathy is a beautiful extension of Ito's stochastic calculus. In fact since Ito's techniques are no more working in the non commutative case, Hudson and Parthasarathy have used other methods. These methods put well in light the structures really involved by Ito's stochastic calculus.

3.3.6 Large deviations theory of Varadhan, D. Stroock, Wentcell-Freidlin...

3.3.7 Measure theory on locally convex spaces [FER 67][PRO 56][SCH 73]...

3.3.8 Semi-martingales theory applied to stochastic control, to stochastic filtering, to stochastic identification

3.3.9 Various complementary theories with respect to semi-martingale theory developping complementary methods for classical physics, mechanics.

3.3.10 The theory of cylindrical distributions.

In general modern stochastic analysis is born coming back to physics and also to Wiener's work and researching the mathematical structures needed there. For example 3.3.7. of [WIE 23] in terms of topological measure theory. Also the fundamental works of I. Segal on Gaussian linear processes and on chaös have similar origins.

4. CONCLUSION

At the Wiener time, and a fortiori today, physics needs much more complex mathematical structures than $(\Omega, \mathfrak{A}, P)$ because physical concepts cannot be implemented only working with abstract probabilized spaces. Also today's infinite dimensional analysis needs complex structures of this type. Hence a corresponding structuration of integration theory is needed. We propose to realize this, starting with set theoretical integration theory and the basic probabilistic concepts formulated in terms of mathematical physics. Then additional structures are progressively added : paradoxes are suppressed in this way and therefore more methods are proposed to physicists : [see KRE 90]

REFERENCES

[BIL 68] P. BILLINSLEY, Convergence of probability measure. John Wiley. New York, (1968).

[BOU 58] N. BOURBAKI, Eléments de mathématique. Topologie Générale. Chapitre 9. Hermann (1958).

[KRE 90] P. KREE, Lecture at the Bielefeld conference on white noise analysis (July 89).

[KRE 89] P. KREE, Comptes Rendus t 308 - Serie 1, pp, 155.158, (1989).

[KOL 33] A.N. KOLMOGOROFF, Grundbegriffe der Wahrscheinlichkeits rechnung. Erg. der Math (Springer editor)Berlin 1933.

[NEL 67] E. NELSON, Dynamical theories of Brownian motions, Mathematical Notes Princeton (1967).

[PRO 56] Y.V. PROHOROV, Convergence of random processes and limit theorems in probability theory. Th. of Pr. and Appl. I,2, (1956), pp, 157.214.

[RAO 88] M.M. RAO, Paradoxes in conditional probability. Journal of Mult. Analysis 27 (1988) pp, 434. 446.

[SCH 73] L. SCHWARTZ, Radon measures on Topological spaces, (1973).

[SIM 74] B. SIMON, The $P(\phi)_2$ Euclidean (Quantum) Field Theory. Princeton University Press (1974).

[STR 88] D. STROOCK, American Scientist., vol 76, n° 4, July-August, (1988).

[WIE 23] N. WIENER, Differential space, J. Math. Phys. 2, (1923), pp, 131.174.

WARD IDENTITIES FOR CONFORMAL MODELS

S. Lazzarini and R. Stora
L.A.P.P.
B.P. 909
74019 Annecy-le-Vieux Cedex
France

The present knowledge concerning the Ward identities which express the symmetry of conformal models is fairly elusive. On the one hand diffeomorphism invariance is invoked, but mostly, locally holomorphic coordinate transformations are used[1][2], in clash with the locality principles of all known versions of quantum field theory. Diffeomorphism invariance is on the other hand understood in terms of Riemannian geometry, but not directly in terms of conformal geometry[3]. Recently, two different sets of Ward identities expressing diffeomorphism invariance in a manifestly conformally invariant way were found for the free bosonic string[4][5]. It turns out that they are equivalent, modulo the free ghost equations of motion. The purpose of this note is to give a geometrical argument showing that the correct invariance for a large class of conformal models, is that of ref. [4], suitably generalized, the simultaneous validity of the invariance depicted in ref. [5] being restricted to the free field situation.

We consider two dimensional actions one may write on an arbitrary compact Riemann surface Σ involving a σ-model type term :

$$S_\sigma = \frac{1}{2} \int_\Sigma \frac{dZ \wedge d\bar{Z}}{2i} \, \partial_Z X^i \, G_{ij}(X) \, \partial_{\bar{Z}} X^j \tag{1}$$

a b-C type term with Thirring type coupling

$$S_{bc} = \int_\Sigma \frac{dZ \wedge d\bar{Z}}{2i} \, b^\alpha \, \partial_{\bar{Z}} C_\alpha + \text{c.c.} +$$

$$\int \frac{dZ \wedge d\bar{Z}}{2i} b^\alpha \, C_\beta \, \bar{b}^{\bar{\alpha}} \, \bar{C}_{\bar{\beta}} \, A_{\alpha\bar{\alpha}}{}^{\beta\bar{\beta}} \tag{2}$$

S. Albeverio et al. (eds.), Stochastics, Algebra and Analysis in Classical and Quantum Dynamics, 169–172.

and, possibly a coupling term of the b-C system with chiral currents constructed with X which exist if the target space coordinated by X has enough symmetries

$$S_{int} = \int_{\Sigma} \frac{dZ \wedge d\bar{Z}}{2i} b\, J_{\bar{Z}}^{(X)}\, C + c.c. \tag{3}$$

The b,C fields are conformal fields of weights (1–j,0), (j,0) respectively and their complex conjugates $\bar{b}$, $\bar{C}$, have weights (0,1–j), (0,j).

These different terms can be expressed in terms of a prescribed set of local analytic coordinates $z, \bar{z}$, in such a way that the dependence of the theory on the complex structure one may define on Σ is described in local terms. Namely, one has

$$dZ = \lambda(dz + \mu d\bar{z}) \tag{4}$$

where μ is a Beltrami differential and λ an integrating factor which fulfills

$$\partial_{\bar{z}}\lambda - \mu\partial_z\lambda = \lambda\partial_z\mu \tag{5}$$

by virtue of $d^2 = 0$.

One then finds:

$$S_\sigma = \frac{1}{2} \int \frac{dz \wedge d\bar{z}}{2i} \frac{1}{1-\mu\bar{\mu}} (\partial_z - \bar{\mu}\partial_{\bar{z}})\, X^i\, G_{ij}(X)\, (\partial_{\bar{z}} - \mu\partial_z)\, X^j \tag{6}$$

$$S_{bc} = \int \frac{dz \wedge d\bar{z}}{2i}\, \beta\, (\partial_{\bar{z}} - \mu\partial_z - j\partial_z\mu)\, \gamma + c.c.$$

$$+ \int \frac{dz \wedge d\bar{z}}{2i} (1 - \mu\bar{\mu})\, \beta^\alpha\, \gamma_\beta\, \bar{\beta}^{\bar{\alpha}}\, \bar{\gamma}_{\bar{\beta}}\, A_{\alpha\bar{\alpha}}{}^{\beta\bar{\beta}} \tag{7}$$

$$S_{int} = \int_{\Sigma} \frac{dz \wedge d\bar{z}}{2i} \beta\, j_{\bar{z}}\, \gamma + c.c. \tag{8}$$

with the definitions

$$b = \frac{\beta}{\lambda^{1-j}} \qquad c = \frac{\gamma}{\lambda^{j}} \qquad J_{\bar{Z}} = \frac{j_{\bar{z}}}{\bar{\lambda}(1-\mu\bar{\mu})} \ . \tag{9}$$

At this point it is clear how to derive the invariance under infinitesimal diffeomorphisms by the replacement

$$(z,\bar{z}) \rightarrow (z + \xi^z \, , \bar{z} + \xi^{\bar{z}}) \tag{10}$$

where $(\xi^z,\xi^{\bar{z}})$ are the coordinates of a smooth vector field.

One thus find (6),(7),(8) to be invariant under the following variations:

$$\begin{aligned}
\delta_\mu &= (\xi.\partial)\mu + \partial_{\bar{z}}\xi^z + \mu\partial_{\bar{z}}\xi^{\bar{z}} - \mu(\partial_z\xi^z + \mu\partial_z\xi^{\bar{z}}) \\
\delta_\beta &= (\xi.\partial)\beta + (1-j)(\partial_z\xi^z + \mu\partial_z\xi^{\bar{z}})\beta \\
\delta\gamma &= (\xi.\partial)\gamma + j(\partial_z\xi^z + \mu\partial_z\xi^{\bar{z}})\gamma \\
\delta X^i &= (\xi.\partial)X^i
\end{aligned} \tag{11}$$

with

$$\xi.\partial = \xi^z\partial_z + \xi^{\bar{z}}\partial_{\bar{z}} \ .$$

The passage to the quantum level requires several comments. First the σ-model has to be proved renormalizable. This is doable when the target space is a homogeneous space[6] and allows the discussion of deformations of (11) other than anomalous dimensions. Besides these possible anomalies one may have a breakdown of conformal invariance associated with a non vanishing of the β functions. If the theory stays conformal, the central charge term in the r.h.s. of the Ward identity corresponding to (11) reads:

$$\int_{\Sigma} \frac{dz \wedge d\bar{z}}{2i} \left[(\Lambda^z\partial_z{}^3\mu - \mu\partial_z{}^3\Lambda^z) + R(\Lambda^z\partial_z\mu - \mu\partial_z\Lambda^z)\right] + c.c. \tag{12}$$

where $\Lambda^z = \lambda^z + \mu\lambda^{\bar{z}}$ and R is a projective connection[7], i.e. in local coordinates z^α, z^β,

$$R^\alpha(dz^\alpha)^2 = R^\beta(dz^\beta)^2 + S(z^\beta; z^\alpha)(dz^\alpha)^2 \tag{13}$$

where $S(z^\beta; z^\alpha)$ is the Schwarzian derivative of z^β with respect to z^α. Additional terms which may break the naïve Ward identity are due to the necessary gauge fixing of global zero modes in the action, when they occur. These may affect the b-c system as well as the σ-model type term. Their study requires a detailed analysis of the geometry.

It is easy to check that solving the Ward identities yields in particular the short distance expansions of operator products involving one - or several energy momentum operators whose correlation functions are obtained by differenciating with respect to μ and $\bar{\mu}$ the vertex functional Γ which extends the classical action at the quantum level.

REFERENCES

[1] A.A. Belavin, A.M. Polyakov, A.B. Zamolodchikov, Nucl. Phys. **B241**, 333 (1984).
[2] D. Friedan, E. Martinec, S. Shenker, Nucl. Phys. **B271**, 93 (1986).
[3] T. Eguchi, H. Ooguri, Nucl. Phys. **B282**, 308 (1987).
[4] C. Becchi, Nucl. Phys. B., to appear.
[5] L. Baulieu, M. Bellon, Phys. Lett. **B196**, 142 (1987).
[6] C. Becchi, A. Blasi, G. Bonneau, R. Collina, F. Delduc, CERN-TH.4996 (1988); A. Blasi, F. Delduc, S.P. Sorella, CERN-TH.5046 (1988); C. Becchi, O. Piguet, CERN-TH.5051 (1988).
[7] R.C. Gunning, Lectures on Riemann surfaces, Princeton Mathematical Notes, Princeton University Press (1966), Princeton N.J., USA.

DYNAMICAL SYSTEMS IN PLASMA THEORY

E.K. MASCHKE
Association EURATOM-CEA sur la Fusion
DRFC - CEN/Cadarache
F-13108 SAINT PAUL LEZ DURANCE, Cédex (FRANCE)

ABSTRACT - After a short introduction into the theoretical description of magnetically confined plasmas, typical dynamical systems describing plasma phenomena are discussed.

1. INTRODUCTION

This paper, which was presented at a meeting on mathematical physics, is intended as a short introduction to situations in plasma physics, which can be described by particular dynamical systems. In Section 2, we briefly recall the basic physical and mathematical concepts used for describing a plasma. In Section 3, we first discuss some general properties of plasmas, which lead to the possibility of describing certain physical phenomena in terms of relatively simple dynamical systems. We then describe typical cases : the magnetic field in a torus, a charged particle in an electric potential wave, and a dissipative system describing nonlinear coupling of waves. In Section 4, we discuss mathematical and numerical results on successive bifurcations in magneto-hydrodynamic problems.

2. THEORETICAL DESCRIPTION OF PLASMAS IN MAGNETIC FIELDS

2.1. PHYSICAL DESCRIPTION OF A PLASMA [1]

A plasma is an ionized gas consisting of several species of particles : electrons (charge $-e$), ions (charge Ze), and eventually neutral atoms. Each species s will be characterized by its local particle density n_s ($s = e$ for electrons, $s = i$ for a single ion species), and by its local temperature T_s if the velocity distribution is locally Maxwellian (otherwise it is necessary to characterize each species by a distribution function f_s, see below). Note that in a plasma, the temperatures of the different particle species are usually not equal (T_e different from T_i). For simplicity we shall however often take $T_e \approx T_i = T$ in the present paper.

S. Albeverio et al. (eds.), Stochastics, Algebra and Analysis in Classical and Quantum Dynamics, 173–187.

The plasma state is obtained when a neutral gas is heated up to sufficient temperatures. The degree of ionization is a step-like function of temperature yielding nearly complete ionization above a density-dependent threshold (Fig. 1). An essential property of a plasma is its capability of screening off electric fields, which leads to local quasi-neutrality and to collective behaviour. This screening is a characteristic feature of the Coulomb force acting between the charged particles. The force field of any particular ion (Fig. 2) is given by the (long-range) Coulomb potential $\phi = e/r$. It attracts a "cloud" of electrons, which screens the field of the ion beyond a distance depending on the electron temperature. This screening distance is called Debye length :

$$\lambda_D = 740\ (T[eV] \ /\ n[cm^{-3}])^{1/2}$$

If this length is larger than the interparticle distance $d = n^{-1/3}$, the plasma has the property of quasi-neutrality ($Zn_i \approx n_e$), and it exhibits collective behaviour, which implies that the plasma is able to propagate and to absorb or amplify many types of waves. A typical example is the plasma wave, which is a longitudinal wave of frequency :

$$\omega_p = (4\pi\ n\ e^2/m)^{1/2}$$

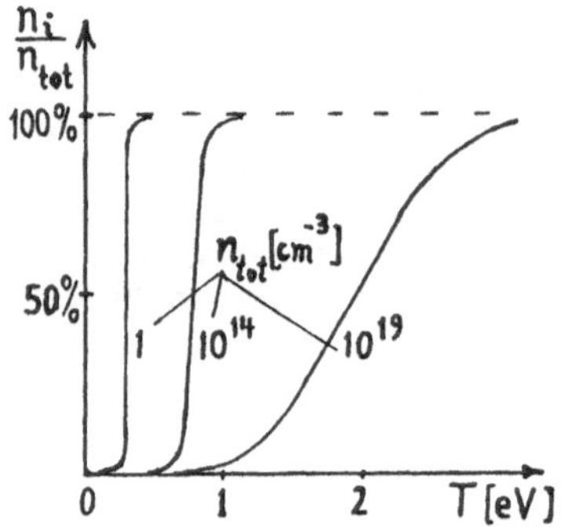

Fig. 1

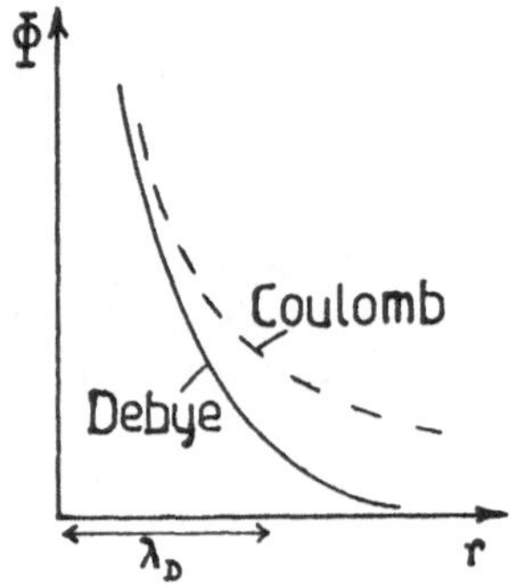

Fig. 2

If the interparticle distance is less than the de Broglie wavelength λ_B of the average particle momentum, ($\lambda_B \approx 1/\sqrt{T}$), that is for $d < \lambda_B$, quantum effects appear. They will not be considered in the present paper.

A survey of plasmas in different regimes is represented schematically in Fig. 3. We shall here limit ourselves to the parameter range of fusion plasmas, that is, T of the order of 10^5 eV (1 eV corresponds to 10^4 K) and n of the order of 10^{14} cm^{-3}.

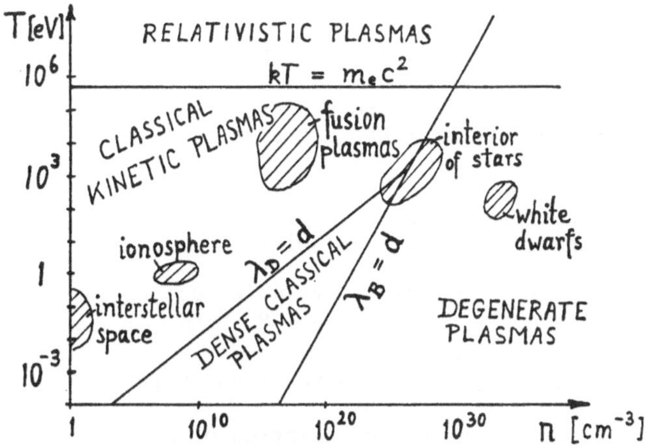

Fig. 3

2.2. MATHEMATICAL DESCRIPTION OF A PLASMA

The general description of a gas of interacting particles requires kinetic equations for distribution functions. In particular, the one-particle distribution function for the particle species s, f_s ($\underline{x}$, $\underline{v}$, t), obeys a kinetic equation of the form :

$$df_s/dt = C_s \quad (= \text{"collision" term})$$

where the right hand side, C_s, expresses the interaction between particles. For details, the reader may consult the general references given at the end of this paper [1-3]. For plasmas immerged in electric and magnetic fields ($\underline{E}$, $\underline{B}$) the kinetic equation takes the form :

$$\frac{\partial f_s}{\partial t} + \sum_{k=1}^{3} v_k \frac{\partial f_s}{\partial x_k} + \frac{1}{m_s} \sum_{k=1}^{3} F_{s,k} \frac{\partial f_s}{\partial v_k} = C_s$$

where $\underline{F}_s$ is the force due to the electromagnetic fields acting on a particle of species s (charge e_s) :

$$\underline{F}_s = e_s \underline{E} + (e_s/c) (\underline{v} \times \underline{B})$$

The electromagnetic field is governed by Maxwell's equations.

By taking moments of the kinetic equations one obtains magneto-hydrodynamic equations for each particle species [4]. Typically, if $n_s(\underline{x}, t)$ denotes the density and $\underline{V}(\underline{x}, t)$ the mean velocity of particle species s, fluid equations of the following form are obtained :

$$m_s n_s (\partial \underline{V}_s/\partial t + \underline{V}_s \cdot \underline{\nabla}\, \underline{V}_s)$$

$$= - \underline{\nabla} \cdot P_s + e_s n_s (\underline{E} + (1/c)(\underline{V}_s \times \underline{B})) + \underline{R}_s$$

Here P_s denotes the complete pressure tensor (including viscosity), and $\underline{R}_s$ is the momentum transfer between species.

A further simplification of the description is obtained by the construction of equations for a single fluid characterized by a mass density ϱ and a mean fluid velocity $\underline{V}$ defined by :

$$\varrho = \sum_s m_s n_s, \qquad \underline{V} = (1/\varrho) \sum_s m_s n_s \underline{V}_s$$

The derivation of such equations can be found in ref. 4.

It is important to note that the description of a plasma by fluid equations neglects certain effects, which are due to the detailed structure of the particle distribution function (e.g. deviations from a locally Maxwellian velocity distribution). However, for plasmas in strong magnetic fields, the magneto-hydrodynamic description yields important information about the stability of large-scale motion of the plasma.

2.3. TOROIDAL PLASMAS FOR NUCLEAR FUSION

In order to confine a plasma in some finite volume by means of a magnetic field, one has to use the properties of the motion of the individual particles and the equilibrium and stability properties of the collective motions of the plasma.

The trajectory of an individual charged particle in a strong magnetic field is a helix which follows approximately a magnetic field line.

More precisely, the motion can be described as the superposition of a rapid rotation (with cyclotron frequency $\omega_c = eB/mc$) about a fictitious point called "guiding center", which itself executes a motion along the magnetic field and slowly drifts across the field if the latter is non-uniform (Fig. 4). If the magnetic field had only a "toroidal" component B_ϕ (in cyclindrical coordinates, see Fig. 4), the confinement of a plasma would not be possible because electrons and ions drift in opposite directions along the axis of such a field (z-axis). The tokamak is a device, in which a toroidal plasma current I_p creates an additional "poloidal" magnetic field $\underline{B}_p$, which together with the "toroidal" field B_ϕ leads to a closed magnetic field configuration as shown in Fig. 4. In such a device, individual particles are confined inside a toroidal volume. But the confinement of the plasma as a whole may be deteriorated by unstable collective motions (plasma instabilities) [1-3].

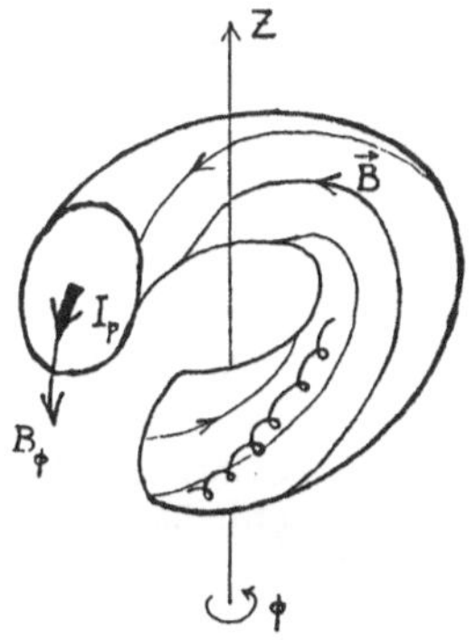

Fig. 4

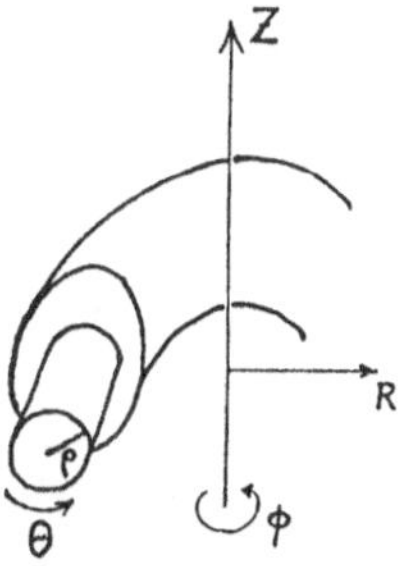

Fig. 5

3. *DYNAMICAL SYSTEMS DESCRIBING PLASMA PHENOMENA*

3.1. GENERAL FEATURES LEADING TO DYNAMICAL SYSTEMS

In the following we shall focus our discussion on toroidal plasmas of the type produced in tokamaks.

In a high-temperature toroidal plasma, there are many possibilities for coupling and resonances between a small number of degrees of freedom of the system. As we shall see in the following, resonances may occur in the magnetic field itself ; there may also be resonances between individual particles and various collective motions (waves) of the plasma ; and there may exist (nonlinear) interactions of a small number of different waves. The presence of such resonant phenomena between a few degrees of freedom of the plasma is the general feature which leads to the possibility of describing such phenomena by relatively simple dynamical systems.

3.2. THE MAGNETIC FIELD IN A TORUS : A HAMILTONIAN SYSTEM

We consider a magnetic field of the type which is created in a tokamak device (see Fig. 4). Using cylindrical coordinates, the equations of magnetic field lines are (assuming $B_\phi \neq 0$ everywhere in the torus) :

$$dR/d\phi = R\, B_R/B_\phi \qquad dZ/d\phi = R\, B_Z/B_\phi$$

and we have

$$\mathrm{div}\, \underline{B} = 0$$

As a consequence of the latter relation, the field line equations may be derived from a Hamiltonian [5-7]. In the case of a magnetic field having a non-vanishing toroidal component, i.e. $B_\phi \neq 0$ in the torus, it is convenient to use again cylindrical coordinates R, ϕ, Z and to interpret ϕ as a time-like coordinate. Instead of R, Z we may consider toroidal coordinates ϱ, θ, where ϱ labels a set of nested tori and θ is a "poloidal" angle as shown in Fig. 5. We assume $(\underline{\nabla}\varrho \times \underline{\nabla}\theta) \cdot \underline{\nabla}\phi \neq 0$ in the region of interest. As shown in ref. 6, the magnetic field can then be written as :

$$\underline{B} = \underline{\nabla}\, \Psi \times \underline{\nabla}\theta \; + \; \underline{\nabla}\phi \times \underline{\nabla}\psi$$

and the field line equations take the form of Hamiltonian equations :

$$d\theta/d\phi = \partial\psi/\partial\,\Psi \qquad\qquad \partial\Psi/\partial\phi = -\,\partial\psi/\partial\theta$$

This shows that we may interpret θ and Ψ as coordinate and canonical momentum, respectively, and ψ as the time-dependent Hamiltonian of a one-dimensional system.

In order to investigate the structure of the magnetic field, a field line passing through a point 0 is followed around the torus, and the successive intersections with a median plane ϕ = const. containing the point 0 are plotted (Fig. 6a). In the case of axisymmetry, the field line Hamiltonian ψ is independant of ϕ, and the field lines lie on so-called magnetic surfaces given by ψ = const.

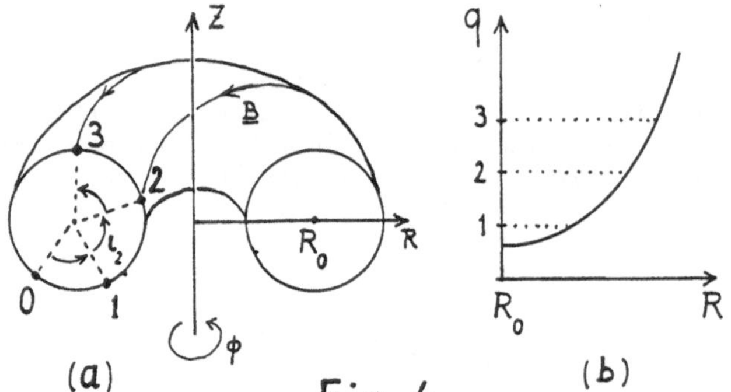

Fig. 6

A winding number can be defined as the limit :

$$1/q = \lim_{N\to\infty} 1/N \sum_{k=1}^{N} \iota_k$$

where ι_k denotes the successive rotation angles shown in Fig. 6a.

In a tokamak, q has a radial dependance as shown in Fig. 6b. Near rational values of q, resonance phenomena may occur, and certain plasma instabilities give rise to perturbations of the magnetic field. In particular, the tearing instability gives rise to a topological change of the magnetic surfaces : for a small helical perturbation $\sim \exp i\,(m\theta - n\phi)$, the magnetic surfaces still exist but "magnetic islands" appear near the "mode-rational" surface where $q = m/n$ (Fig. 7a). For a sufficiently strong perturbation, the region near the separatix S of Fig. 7a becomes stochastic due to the interaction of the helical perturbation with the Fourier components of the θ-variation of the magnetic field in the torus. In the case of a tearing instability near $q = 1$, theoretical studies [8] and numerical calculations [9] have allowed to determine the width of the stochastic (or turbulent) region as a function of the amplitude of the perturbation (Fig. 7b). This case is typical of the kind of problems encountered in the physics of magnetically confined plasmas.

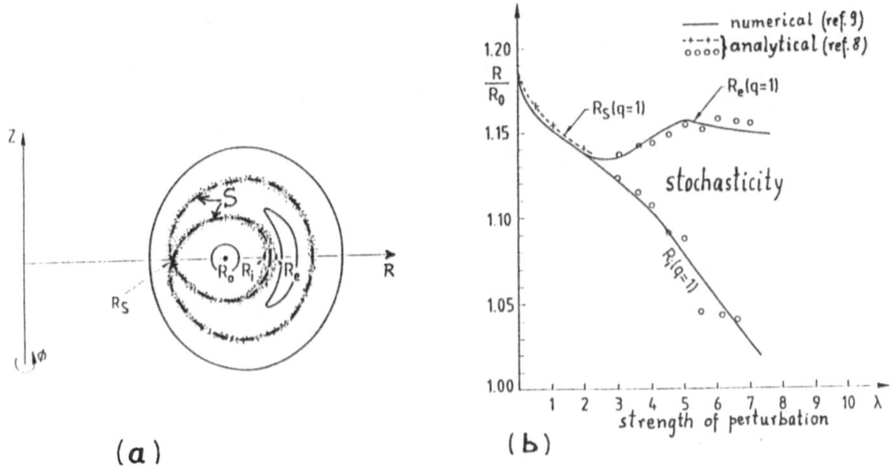

Fig. 7

3.3. CHARGED PARTICLE IN AN ELECTRIC POTENTIAL WAVE

As said before, a plasma is able to propagate, absorb or amplify many types of waves. An important case is the plasma oscillation (or Langmuir wave). It is characterized by an electric field which can be derived from an electric potential $U \sim \exp\{i\,(kx - \omega t)\}$. An electron moving with a velocity V along the x-axis and interacting with such a potential is described by a Hamiltonian of the following type (using appropriate normalization of length and time) :

$$H = \frac{1}{2} V^2 - P \cos K\,(x - t)$$

If there is more than one wave we have to add the corresponding terms in the potential energy. For instance, in the coordinate frame moving with the first wave, we may have :

$$H = \frac{1}{2} V^2 - P_1 \cos x - P_2 \cos K (x - t)$$

This is the simplest non-integrable Hamiltonian. It has been extensively studied, and we shall not discuss it here but refer the interested reader to a review article [10].

3.4. NONLINEAR COUPLING OF WAVES : A DISSIPATIVE SYSTEM

If different waves can occur simultaneously in a plasma, they often influence each other. Such interactions of waves are described by coupling terms involving the amplitudes of two or more waves. The simplest nontrivial dynamical system describing such wave coupling is obtained when we consider the following physical situation [11].

We assume that there is a high-frequency wave, which is unstable according to linear perturbation theory. This wave is characterized by its complex amplitude $A_0 = |A_0| \exp(i\phi_0)$, its real frequency ω_0 and its growth rate $\gamma_0 > 0$. This wave is coupled to a damped wave of lower frequency, which is characterized by an amplitude $A_1 = |A_1| \exp(i\phi_1)$, its real frequency ω_1 and a (negative) linear growth rate $\gamma_1 < 0$. Assuming a relatively simple nonlinear coupling of these waves one arrives at the following model equations :

$$i \, (dA_0/dt - \gamma_0 A_0) = V A_1^2 \, e^{i\Delta\omega t}$$

$$i \, (dA_1/dt - \gamma_1 A_1) = V A_0 A_1^* \, e^{-i\Delta\omega t}$$

where $\Delta\omega = \omega_0 - 2\,\omega_1$, and V is a real coupling coefficient. Changing variables according to :

$$x = (V/\gamma_0) \; |A_0| \sin\theta, \qquad y = (V/\gamma_0) \; |A_0| \cos\theta$$

$$z = (V/\gamma_0)^2 \; |A_1|^2 , \qquad \tau = \gamma_0 t$$

one obtains the folling dissipative dynamical system :

$$dx/dt = x + \delta y - z + 2y^2$$

$$dy/dt = y - \delta x - 2xy$$

$$dz/dt = - 2\gamma z + 2xz$$

where $\gamma = - \gamma_1/\gamma_0$, $\delta = \Delta\omega/\gamma_0$, and $z > 0$.

The system has two fixed points,
$\{0, 0, 0\}$, and $C = \{\gamma, \gamma/(1-2\gamma), \gamma[1 + \sigma^2/(1 - 2\gamma)^2]\}$

There exist extensive theoretical and numerical studies of the attractors of the above system [11]. When the parameters γ and σ are varied, attractors undergo bifurcations of different types. Fig. 8 shows that there are three regions with different behaviour. An interesting feature is the occurrence of intermittency in a transition from a limit cycle to a strange attractor. Such a behaviour may be observable in certain plasma experiments.

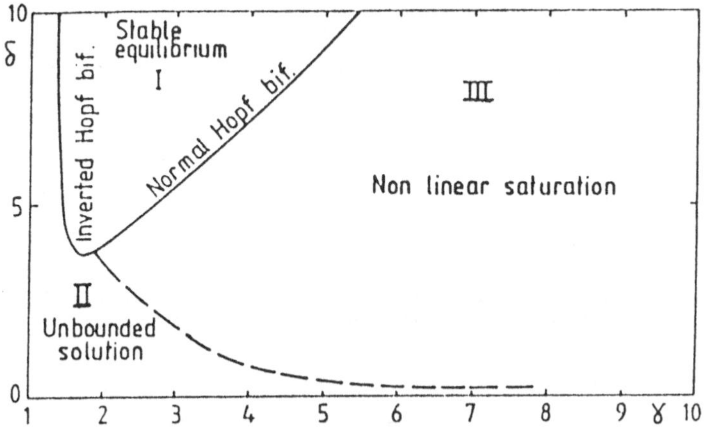

Fig. 8. Three regions in (γ,δ)-space [from ref. 11]

4. SUCCESSIVE BIFURCATIONS IN A 2D CONVECTION PROBLEM

4.1. FORMAL EQUIVALENCE BETWEEN THE 2D BENARD CONVECTION AND CERTAIN MAGNETO-HYDRODYNAMIC INSTABILITIES

In the presence of a strong magnetic field, the unstable motion of a plasma is often nearly two-dimensional, and in fact can be described approximately by equations which formally ressemble those of a fluid layer heated from below (Benard convection problem). As shown in ref. 12, this is true in particular for an interchange-unstable plasma layer in a curved magnetic field. Let us consider the case of a plasma situated between two concentric cylinders which are kept at two different temperatures, T_0 and T_1, with $T_0 > T_1$. The magnetic field is purely azimuthal, $\underline{B} = B(\underline{x})\ \underline{e}_\theta$. The plasma is described by magneto-hydrodynamic (MHD) equations:

$$\rho\ (\partial V/\partial t + \underline{V}.\underline{\nabla V}) = -\ \underline{\nabla}p + (\underline{\nabla} \times \underline{B}) \times \underline{B} + \mu\Delta\underline{V}$$

$$\partial\rho/\partial t + \underline{\nabla}.(\rho\underline{V}) = 0$$

$$\partial\underline{B}/\partial t = \underline{\nabla} \times (\underline{V} \times \underline{B}) - \eta\ \underline{\nabla} \times (\underline{\nabla} \times \underline{B})$$

$$C\rho\ (\partial T/dt + \underline{V}.\underline{\nabla}T) + p\ \underline{\nabla}.\underline{V} = K\ \Delta\ T$$

$$\underline{\nabla}\ .\ \underline{B} = 0, \qquad p = p\ (\rho,\ T)$$

where η is the electric resistivity, μ the viscosity, K the thermal conductivity. This complicated system of equations can be considerably simplified if we make the following assumptions :

- the motion is two-dimensional, hence :
  $\partial/\partial\theta = 0, \quad V_\theta = 0, \quad \underline{B} = B(R, Z)\,\underline{e}_\theta$
- weak curvature of the cylindrical boundaries $R = R_0 \pm d/2$, i.e. $R_0/d \gg 1$
- plasma pressure small compared to magnetic field pressure, i.e. $\beta = p/B^2 \ll 1$
- nearly constant density, $\varrho \approx \varrho_0$, and $p = \text{const.}\ T$.

With these assumptions, the velocity field becomes nearly divergence-free and can be represented, to lowest order in d/R_0, by a stream function ϕ :

$$\underline{V} = \underline{e}_\theta \times \underline{\nabla}\phi(R, Z)$$

Using the above assumptions it is then possible to reduce the full system of MHD equations to a simplified system of equations for the two unknowns ϕ and $\tilde{T} = T - T_{eq}(R)$ (where $T_{eq}(R)$ is the equilibrium temperature in the absence of motion). Introducing the variables $x = (R - R_0)/d$, $y = Z/d$, these simplified equations take the following form :

$$\partial\Delta\phi/\partial t + 1/\sigma\,[(\partial\phi/\partial x)(\partial\Delta\phi/\partial y) - (\partial\phi/\partial y)(\partial\phi/\partial x)] = R\,\partial\tilde{T}/\partial y + \Delta^2\phi$$

$$\partial\tilde{T}/\partial t + 1/\sigma\,[(\partial\phi/\partial x)(\partial\tilde{T}/\partial y) - (\partial\phi/\partial y)(\partial\tilde{T}/\partial x)] = 1/\sigma\,\Delta\tilde{T} + 1/\sigma\,[T'_{eq}/(T_0 - T_1)]\,\partial\phi/\partial y$$

Here $\sigma = (\mu/\varrho_0)K$ is the Prandtl number, and

$$\mathbf{R} = [2d/R_0]\,[(T_0 - T_1)/T_0]\,[d^2/(K\mu/\varrho_0^2)]$$

is the analogue of the Rayleigh number. The system of equations thus obtained is identical with the equations for two-dimensional(2D) Benard convection. Therefore, it is interesting to study 2D Benard convection as a theoretical model for the motion of a plasma in a strong magnetic field. In the following we shall describe a systematic mathematical and numerical investigation of this model problem [12].

4.2. SPECTRAL REPRESENTATION AND TRUNCATION

The above equations for 2D Benard convection have been studied for boundary conditions of periodicity in y, and $\phi = \Delta\phi = 0$, $\tilde{T} = 0$ for $x = 0, 1$. The unknowns ϕ, $\tilde{T}$ are written in Fourier series :

$$\phi(x, y, t) = \sum_{m,n} \phi_{mn}(t) \sin(m\pi x) \sin(2n\pi y/L)$$

$$\tilde{T}(x, y, t) = \sum_{m,n} \tilde{T}_{mn}(t) \sin(m\pi x) \cos(2n\pi y/L)$$

The investigation of ref. 12 had been deliberately restricted to the subspace of solutions such that (m + n) is even (the equations are invariant with respect to the corresponding symmetry).

The solutions are then approximated by the following truncation (Galerkin approximation) :

$$m + n \leqslant 2p \qquad (p = 1, 2, 3, \ldots)$$

For a truncation with given p, one obtains a system of $N = 2p^2 + p$ ordinary nonlinear differential equations for the amplitudes(*) $\phi_{mn}(t)$, $T_{mn}(t)$. For p = 1, a system of three equations is obtained, which is equivalent to the well-known Lorenz equations. The sequence of increasing number p corresponds to a sequence of dynamical systems with increasing number of degrees of freedom $N = 2p^2 + p$.

Since our original problem of Benard convection is described by partial differential equations, i.e., an infinite number of degrees of freedom, the question arises :
Is it possible to approximate the attracting solutions of the partial differential equations by the attracting solutions of a dynamical system obtained by the above truncation and how many terms should be retained ?

This question has lead to a number of mathematical and numerical studies, which we shall discuss in the following sections.

4.3. THE SUFFICIENT NUMBER OF "MODES"

In relation to the question posed at the end of the preceding paragraph, a fundamental mathematical contribution has been made by FOIAS and PRODI [13]. They considered a fluid equation of motion with a given external force (instead of the temperature-dependent boyancy force), and with "rigid" boundary conditions ($\underline{V} = 0$ at the boundaries). The coupled

---

(*) - Although from the point of view of the physicist, the number of different couples (m, n) retained in the truncation should be called the "number of modes", it is usual in fluid dynamics to call the number N the number of retained modes. We shall here refer to N as the number of degrees of freedom of the truncated system.

system of fluid equation and heat transport equation including "free slip" boundary conditions (Section 4.1) has been treated by SARAMITO [14, 15] and FOIAS et al. [16]. For the purpose of the present survey we may summarize the results of these authors in terms of our notations as follows.
Let the couples (m, n) be ordered in a sequence corresponding to increasing eigenvalues of the Laplacian. Thus, the eigenvalues :

$$\lambda_{mn} = 2\pi^2 [m^2 + (2\pi n/L)^2]$$

are ordered in an increasing sequence and relabeled ($\lambda_{mn} \Longrightarrow \lambda_k$) :

$$0 < \lambda_1 \leqslant \lambda_2 \leqslant \ldots \leqslant \lambda_k \leqslant \ldots$$

We then have a one-to-one correspondence between the couples (m, n) and the positive integers. Let P_K {Φ, T} be the projection of {Φ, T} onto the linear space of dimension K defined by the first K basis functions corresponding to the above ordered sequence. We then have the following :

<u>Theorem</u> : If for $t \to \infty$ the projection P_K {Φ, $\ddot{T}$} has a definite time behaviour, which may be stationary, periodic, quasi-periodic, ... ; then the solution {Φ, $\tilde{T}$} has the same asymptotic time behaviour if the quantity $\mu_k = \inf_{k>K} \{\lambda_k\}$ satisfies a sufficient criterion which may be written :

$$\mu_k - (R/\sigma) A_j \mu_K^{-1/2j} - R^{1/2} [1 + \mu_K^{1/2}]^{1/2} > 0$$

Here j is a positive integer, which must be chosen to minimize the left hand side of the inequality, and :

$$A_j = 1/\pi\sqrt{2} (j!\ a_0\ L^{1/2})^{1/j} \qquad (j = 1, 2, 3, \ldots)$$

$$a_0 = [2 + \sqrt{2}/\pi + (2\pi)^{-2}]^{1/2} \approx 1.5$$

<u>Remark</u> : Minimization with respect to j for large j using Stirling's formula [16] yields an approximate criterion for $(R/\sqrt{\mu_K}) \ll 1$:

$$\mu_K > (R/\sigma) \log(C\ R/\sigma)$$

For the truncation procedure introduced in the preceding section, we have for given $p \gg 1$ the approximate relation :

$$\mu_K \approx \pi^2 N = \pi^2 (2p^2 + p)$$

which allows an easy evaluation of the above criterion.

4.4. SUCCESSIVE BIFURCATIONS

The truncation procedure of Section 4.1 leads to a sequence of dynamical systems with $N = 2p^2 + p$ degrees of freedom (p = 1, 2, ...). Considering the Rayleigh number R as the bifurcation parameter and fixing the Prandtl number, $\sigma = 10$, a numerical investigation of systems with values of p up to 15 (hence up to N = 465) has been made [12, 15]. The result is shown in Fig. 9, where for each system with given p or N (abscissa), the Rayleigh number for bifurcation to a new solution branch is plotted (ordinate : $r = \sqrt{R/R_c}$, with $R_c = 658$ the critical Rayleigh number for onset of stationary convection ; the points r = 1 corresponding to this onset have been omitted).

For definiteness, let us consider the system with N = 55 (p = 5). With increasing r > 1, the stationary convective state first undergoes a Hopf bifurcation (indicated by a cross) then a transition to a strange attractor of the Lorenz type (indicated by the letter A). For higher r the system returns to a periodic behaviour (letter A) until a period doubling bifurcation is reached (indicated by a double cross), followed by a bifurcation to a torus (letter T). As can be seen on Fig. 9, the values of r for the different bifurcations converge as N is increased, and the solutions of the strange attractor type disappear for large N. Note that the investigation was limited to a single value of the Prandtl number σ, and that the disappearence of the strange attractors for large N may not be a general phenomenon.

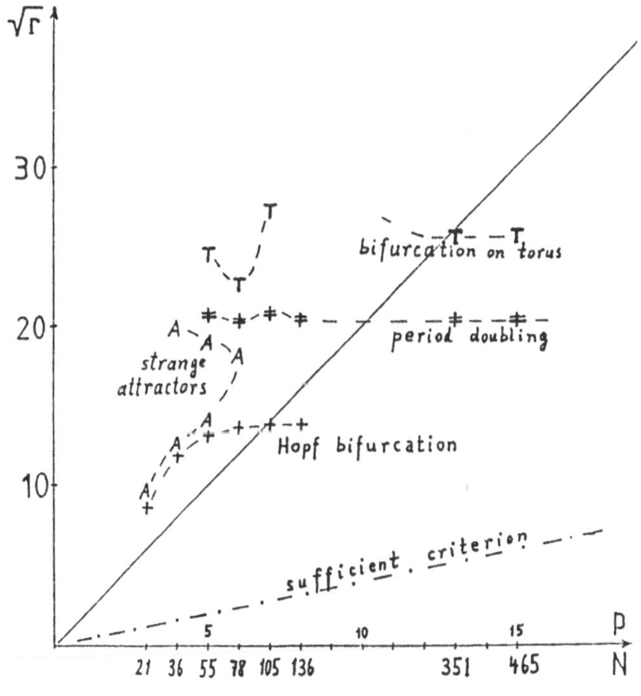

Fig. 9

In addition to the numerical results, Fig. 9 also shows the limit given by the sufficient criterion discussed in Section 4.3. It must be stressed that this criterion does not deal with the problem of convergence, but concerns a statement about the behaviour of a projection of the exact solution.

CONCLUSION

The cases discussed in this paper have been chosen in order to illustrate by a few typical examples the way in which particular dynamical systems arise in the theoretical description of plasma phenomena. The present paper does not attempt to give a complete picture of the many investigations published in the litterature, but is intended as a simple introduction for non-specialists.

ACKNOWLEDGEMENT

I would like to thank Anne Bertin-Maghit for her efficient help in preparing and arranging the typescript of this paper.

REFERENCES

[1] - F. CHEN, Introduction to Plasma Physics, Plenum Press, New York 1976.

[2] - M.A. LEONTOVICH, Ed., Reviews in Plasma Physics, Consultants Bureau, New York.

[3] - M.N. ROSENBLUTH and R.Z. SAGDEEV, Eds., Handbook of Plasma Physics, Vol. 1 and 2, North Holland Publishing Company, Amsterdam 1983.

[4] - S.I. BRAGINSKII, in ref. [2] - Vol. 1 (1965) p 205.

[5] - K.J. WHITEMAN, Rep. Prog. Phys. 40 (1977) 1033.

[6] - A.K. BOOZER, Phys. Fluids 26 (1983) 1288 also, report Plasma Physics Lab. Princeton PPPL-2094 (1984).

[7] - K. ELSASSER, Plasma Phys. Contr. Fusion 28 (1986) 1743.

[8] - A.J. LICHTENBERG, Nucl. Fusion 24 (1984) 1277.

[9] - C. MERCIER, Sov. J. Plasma Phys. 9 (1983) 82.

[10] - D. ESCANDE, Physics Reports 121 (n° 3, 4) (1985).

[11] - C. MEUNIER, M.N. BUSSAC and G. LAVAL, Physica 4D (1982) 236.

[12] - E.K. MASCHKE and B. SARAMITO Physica Scripta Vol. T2/2 (1982) 410.

[13] - C. FOIAS and G. PRODI, Rend. Semin Mat. Univ. Padova - 39 (1967) 1.

[14] - B. SARAMITO, Report EUR-CEA-FC-1125 (Fontenay-aux-Roses, France) (1981).

[15] - E.K. MASCHKE and B. SARAMITO, Phys. Lett. 88A (1982) 154.

[16] - C. FOIAS, O.P. MANLEY, R. TEMAM, Y.M. TREVE Phys. Rev. Lett. 50 (1983) 1031.

BOSONIC STRINGS AND MEASURES ON INFINITE DIMENSIONAL MANIFOLDS

S. PAYCHA

based on joint work with
S. ALBEVERIO, R. HOEGH - KROHN, S. SCARLATTI

dedicated to the memory of Raphael Höegh.Krohn

Mathematisches Institut
Lehrstuhl Prof. Albeverio
Ruhr Universität Bochum
4630 Bochum FRG

A B S T R A C T

This paper reports and extends the results of [1] in which a correspondence was set up between the Polyakov model for dimension of space - time $d \leq 12$ and the Liouville model studied in [25]. With the aim of further improving the mathematical framework underlying the reduction of the Polyakov measure to the Liouville measure already set up in [1], it gives a precise description of the formal steps of the reduction procedure in the context of Gaussian measures on infinite dimensional manifolds.

S. Albeverio et al. (eds.), Stochastics, Algebra and Analysis in Classical and Quantum Dynamics, 189–203.

Introduction

In [1], we set up a correspondence between the quantization of closed bosonic strings for space - time dimension $d \leq 12$ with the quantization of the classical Liouville model corresponding to a free massless field in two dimensions perturbed by an exponential potential. This correspondence arises from an interpretation of the formal "Polyakov" measure that describes the functional quantization of bosonic strings evolving in a space - time of dimension $d \leq 12$ in terms of the Liouville measure (which describes the quantization of the classical Liouville model) coupled with an integration over a finite dimensional manifold, namely the Teichmüller space of the surfaces describing the classical motion of a closed bosonic string. This upper bound for the space-time dimension, (which can infact be chosen larger -$d <$ 20-private communication S. Kusuoka) is dictated by the singularities of the two - dimensional Greens function. This relation between the two models is established along the lines of the usual procedure for the functional quantization of bosonic strings ([2],[3],[4],[5]) and involves formal measures on infinite dimensional manifolds, in particular on the manifold of smooth Riemannian metrics on a two dimensional Riemannian surface. In analogy to the finite dimensional case, one is led to a formal identification of the measures on the infinite dimensional manifolds with measures on the tangent spaces to these manifolds. The transformation of measures required in the Polyakov procedure are then performed on the level of the tangent spaces.

In this paper, we set up a mathematical framework in which the formal description of the measures and the formal transformations they undergo in the correspondance between the two models are given a precise meaning. This mathematical description is based on the construction of Gaussian measures on infinite dimensional manifolds made in [6],[7],[8] ... and the study of the manifolds of Riemannian metrics undertaken in [9],[10].

In the first chapter, we recall the general lines of the Polyakov euclidean quantization procedure for closed bosonic strings with genus $p > 1$.

In the second chapter (which is based on a more detailed paper -see [1])), we shall interpret this procedure in terms of local Gaussian measures on infinite dimensional manifolds. In particular, we shall give a meaning to the formal identification of the measures on the infinite dimensional manifolds with measures on tangent spaces to these manifolds and describe the transformation of these formal measures.

Finally, in the third chapter,on the grounds of this mathematical construction and along the lines of a joint work with S. Albeverio, R. Hoegh - Krohn and S. Scarlatti, we shall discuss the correspondence between the functional quantization of closed bosonic strings and the Liouville model.

1. Euclidean quantization of closed bosonic strings; the Polyakov measure

Let us first desribe the classical motion of a bosonic string.

1.1 Classical closed bosonic strings

A closed string will be represented by a smooth map from the unit sphere S^1 in $\mathbb{R}^2$ into the d dimensional space $\mathbb{R}^d$ ($d \geq 2$), the euclidean version of space - time.

The motion of the free string evolving in $\mathbb{R}^d$ from time t_1 to time t_2, $t_2 > t_1$ is then naturally described by a smooth map from $S^1 \times [t_1, t_2]$ into $\mathbb{R}^d$. This picture generalizes to interacting strings whose motion are represented by a two dimensional smooth surface $\tilde{\Lambda}$ of genus p embedded in $\mathbb{R}^d$ by a smooth map $x \equiv (x^\mu,\ \mu = 1, \dots d)$, whereby $\tilde{\Lambda}$ is homeomorphic to a smooth compact boundaryless 2 - dimensional surface Λ from which a finite number of points have been removed; the removed points correspond to the incoming and outgoing strings [15].

As in the case of point particles, the motion of the string is given by a minimal action principle. The natural generalization to the string picture of the point particle action given by the length of the curve described by the point particle leads to the choice of an action equal to the area of the surface described by the embedded string in $\mathbb{R}^d$, this surface being equipped with the metric induced by the euclidean metric in $\mathbb{R}^d$, [12]. A minimization of this action w.r.t. the embedding variable x yields a harmonic equation:

$$-\Delta_h x = 0 \tag{1.1}$$

with boundary conditions on x. Here $-\Delta_h$ denotes the Laplace-Beltrami operator with respect to the metric h induced on the embedded surface $x(\tilde{\Lambda})$ by the eucledian metric on $\mathbb{R}^d$.

This equation of motion can also be derived from a minimization of this same action in which the fixed metric h has been replaced by a variable smooth Riemannian metric g on $\tilde{\Lambda}$, the minimization is then done w.r.t. both variables x and g [13]. This generalized action has the form:

$$\tilde{A}(x,g) = \frac{1}{2}\int_{\tilde{\Lambda}} \sqrt{g(\eta)}\, g^{ab}(\eta)\partial_a x^\mu(\eta)\partial_b x^\mu(\eta) d\mu \tag{1.2}$$

Here $\eta = (\eta^1, \eta^2)$ is a smooth real atlas on $\tilde{\Lambda}$.For $a = 1,2$, ∂_a refers to the partialderivative w.r.t the coordinate η^a. $(g^{ab})_{\substack{a=1,2\\ b=1,2}}$ denotes the inverse of the matrix $(g_{ab})_{\substack{a=1,2\\ b=1,2}}$ corresponding to the two covariant tensor describing the matrix g. The root of the determinant of g is denoted by $\sqrt{g}$.

The action (1.2) is clearly invariant under conformal transformations or Weyl transformations:

$$g \longrightarrow e^\varphi g \tag{1.3}$$

where φ is a smooth valued function on $\tilde{\Lambda}$ and $(e^\varphi g)(\eta) \equiv e^{\varphi(\eta)} g(\eta) \quad \forall\, \eta \in \tilde{\Lambda}$.

This action is moreover physically relevant since it is independent of the parametrization of $\tilde{\Lambda}$ in the following sense. For a smooth diffeomorphism f on $\tilde{\Lambda}$, we denote by $f^* \cdot g$ the pull-back of g under the diffeomorphism f, we have:

$$\tilde{A}(x \circ f, f^* \cdot g) = \tilde{A}(x, g). \tag{1.4}$$

as can easily be checked on the formula (1.2).

The two fundamental invariance properties combined with assumptions on the behaviour of the metric g and the embedding x at the punctures of $\tilde{\Lambda}$ enable to extend the domain of integration $\tilde{\Lambda}$ in the action (1.2) to the compactified surface Λ obtained from $\tilde{\Lambda}$ by adding a finite number of points [14]. The symbol Λ will now stand for a 2 dimensional real compact boundaryless manifold with genus $p > 1$.

We introduce the following notations:

$$X \equiv \{ \text{ smooth embeddings of } \Lambda \text{ in } I\!R^d\} \tag{1.5}$$

$$M \equiv \{ \text{ smooth Riemannian metrics on } \Lambda\} \tag{1.6}$$

The action $\tilde{A}$ thus formally extends to an action A on $X \times M$:

$$A(x,g) \equiv \frac{1}{2}\int_\Lambda \sqrt{g(\eta)}g^{ab}(\eta)\partial_a x^\mu(\eta)\partial_b x^\mu(\eta)d\eta \tag{1.7}$$

The invariance properties (1.3) and (1.4) correspond to the invariance of (1.7) under the action of the groups:

$$P \equiv \{e^\varphi, \varphi \in C^\infty(\Lambda, I\!R)\} \tag{1.8}$$

$$D \equiv \{ \text{ smooth diffeomorphisms of } \Lambda\} \tag{1.9}$$

P acting g on M by pointwise multiplication and D acting on M by pull-back.

As we shall point out in the following pragraph, although the invariance under the action of the diffeomorphisms is naturally preserved after quantization, an anomaly refered to as the conformal anomaly will arise from the breaking of the invariance under the conformal transformations after quantization.

Let us briefly describe the quantization procedure.

1.2 Euclidean quantization of closed bosonic strings

Following the general functional quantization procedure, the action (1.7) depending on two variables x and g, the quantization of closed bosonic strings requires the construction of a measure on $X \times M$ of the form:

$$d\mu_p(x,g) \equiv e^{-A(x,g)} D_g[x] D[g] \tag{1.10}$$

where $D_g[x]$ is a heuristic Lebesgue measure on the infinite dimensional space X depending on the metric g and $D[g]$ is a heuristic Lebesgue measure on the infinite dimensional space M. For the sake of simplicity,we have left out the cosmological term $e^{-\mu^2 \int_\Lambda \sqrt{g(\eta)} d\eta}$ usually introduced for renormalization purposes, $\mu \in \mathbb{R}^*/\{0\}$.We shall however take it into account later in this paper. These measures can formally be interpreted as underlying Lebesgue measures for some Gaussian measures on some infinite dimensional manifolds with covariance given by a Riemannian (possibly weak) structure on these manifolds. The fact that the scalar products defining the required Riemannian structures cannot be chosen to be invariant under diffeomorphisms as well as conformally invariant, leads to a similar conclusion concerning the measures namely that measures chosen to be invariant under reparametrizations of the surface Λ can not be invariant under conformal transformations. This non invariance under conformal transformations is then reflected onto the measure (1.10) and accounts for the breaking of conformal invariance mentioned earlier on.

Let us now describe the relevant quantities for the quantized strings namely the scattering amplitudes for m closed incoming and outgoing strings P_j, $j = 1, \ldots, m$ with momenta k_j, $j = 1, \ldots, m$ which can formally be desribed in terms of expectation values w.r.t. the Polyakov measure (1.10) as follows (see e.g. [14],[15]):

$$\int_\Lambda \prod_{j=1}^m d\eta_j \int_M \int \prod_{j=1}^m \left(\sqrt{g(\eta_j)} e^{i(x, y_g^j)_g^d} S_j(x, g) \right) d\mu_P(x, g) \tag{1.11}$$

where y_g^j, $j = 1, \ldots, m$ are $\mathbb{R}^d$ valued distributions on Λ given by:

$$y_g^j \equiv (k^j)^\mu \, y_g^\mu(\cdot, \eta_j) \text{ for fixed } \eta_j \in \Lambda, \; k^j \in \mathbb{R}^d, \; j = 1, \ldots, m$$

and

$$\int_\Lambda \sqrt{g(\eta')} y_g^\mu(\eta, \eta') \lambda(\eta') d\eta' = \lambda(\eta)$$

for all λ in $C^\infty(\Lambda, \mathbb{R})$, the space of smooth $\mathbb{R}$ - valued functions on Λ with $\mu = 1, \ldots, d$. The factors $S_j(x, g)$, $j = 1, \ldots, m$ are functionals on $X \times M$ standing for the contribution of the vertex operators (see e.g. [14],[15]). Here, $(\cdot, \cdot)_g^d$ is a scalar product on X which we shall describe more precisely in the next chapter.

In the specific case of scattering of tachyons [15], we have $S_j = 1$, $j = 1, \ldots, m$ and the scattering amplitudes can therefore, up to a regularization of y_g, be interpreted in terms of coupled Fourier transforms along the space of embeddings combined with an integration against the metrics. In this paper, we shall restrict ourselves to this simpler case.

Let us remark that in both expressions (1.10) and (1.11), the formal Lebesgue measure $D_g[x]$ is tempered by the action term so that $e^{-A(x,g)} D_g[x]$ can in fact be interpreted as a Gaussian measure up to a normalisation term.

The object of the following chapter is to reinterpret the expression (1.11) substituting to $S_j(x, g)$ an approximated contribution $S_j^N(x, g)$ with $\lim_{N \to +\infty} S_j^N(x, g) = S_j(x, g)$, the approximation introducing a tempering term for the formal measure $D[g]$ so that as for the embeddings, we can describe the integration over M in (1.11) in terms of a Gaussian measure on M.

2. A local regularized pathspace measure for closed bosonic strings

In order to describe the expectation (1.11) in terms of Gaussian measures, we first introduce approximated contributions $S_j^N(x,g)$ for the vertex operators, $N > 0$. For a fixed metric g in M and $N > 0$ in $\mathbb{N}$, we define for a matric $\tilde{g}$ in M:

$$S_j^N(x,\tilde{g}) \equiv e^{-\frac{1}{2N}<\tilde{g},\tilde{g}>_s^g} S_j(x,g) \tag{2.1}$$

where $< \cdot,\cdot >_s^g$ is a scalar product of Sobolev class $s > 1$ on M relative to the metric g given by:

$$< h,k >_s^g \equiv \sum_{r=0}^{s} < (\nabla^g)^{(r)} h, (\nabla^g)^{(r)} k >_0^g \tag{2.2}$$

and for two C^∞ tensor fields $T = T_{k,\dots,l}^{i,\dots,j}$ and $T' \equiv T_{k',\dots,l'}^{\prime i',\dots,j'}$ on Λ we have set:

$$< T,T' >_0^g \equiv \int_\Lambda d\eta \sqrt{g(\eta)} g g_{ii'} \dots g_{jj'} g^{kk'} \dots g^{ll'} T_{k,\dots,l}^{i,\dots,j} T_{k',\dots,l'}^{i',\dots,j'} \tag{2.3}$$

Here ∇^g is the Riemannian connection associated to the metric g on $C^\infty(S^2T^*)$ the space of smooth sections of the bundle of symmetric 2 - covariant tensors on Λ, denoted by S^2T^*. Since $\lim_{N\to+\infty} S_j^N(x,\tilde{g}) = S_j(x,\tilde{g})$, the expectation (1.11) in which $S_j(x,g)$ has been replaced by $S_j^N(x,g)$, $j = 1,\dots,m$, defines a natural approximation for the heuristic expectation (1.11) given by:

$$\begin{aligned} &\int_\Lambda \prod_{j=1}^m d\eta_j \int_M \int_X \prod_{j=1}^m \left(\sqrt{g(\eta_j)} e^{i(x,y_g^j)_g^d} S_j^N(x,g) \right) d\mu_p(x,g) \\ &= \int_\Lambda \prod_{j=1}^m d\eta_j \int_M \prod_{j=1}^m \sqrt{\tilde{g}(\eta_j)} e^{-\frac{1}{2N}<\tilde{g},\tilde{g}>_s^g} D[\tilde{g}] \cdot \\ &\int_X \prod_{j=1}^m e^{i(x,y)_{\tilde{g}}^d} e^{-A(x,\tilde{g})} S_j(x,\tilde{g}) D_{\tilde{g}}[x] \end{aligned} \tag{2.4}$$

The formal Lebesgue measure $D[\tilde{g}]$ has thus been tempered by the exponential term $e^{-\frac{1}{2N}<\tilde{g},\tilde{g}>_s^g}$ and yields a formal Gaussian measure (up to a normalization term) with covariance given by a Sobolev scalar product. Let us remark that a metric g has here been specified. The object of the next paragraph is to give a precise description of this formal Gaussian measure which justifies a posteriori the choice of a Sobolev scalar product for the covariance.

2.1 A local measure on the manifold of Riemannian metrics of Sobolev class

Since the space M has the structure of a Fréchet manifold (see e.g. [9]), one procedure to define such measures, namely the one consisting of carrying back onto the manifold through the exponential map a Gaussian measure constructed on the tangent space, does not aply to M. Indeed, the implicit function theorem underlying the existence of an exponential map does not hold on Fréchet manifolds [16]. We shall therefore extend the manifold M to a Hilbert manifold, namely the manifold M^s of Riemannian metrics of Sobolev class $s > 1$:

$$M^s \equiv H^s(S^2T^*) \cap M^c \tag{2.5}$$

whereby M^c is the space of continuous sections of the bundle S^2T^* which induce a scalar product on the tangent space $T\Lambda$ to Λ and $H^s(S^2T^*)$ is the closure of $C^\infty(S^2T^*)$ for the scalar product $< \cdot,\cdot >^g_s$. Let us point out that although the scalar product depends on g, the space $H^s(S^2T^*)$ is independent of the choice of the metric.

The construction of a Gaussian measure on M^s thus boils down to the construction of a Gaussian measure on the tangent spaces to M^s denoted by T_gM^s, $g \in M^s$ and which are C^∞ diffeomorphic to $H^s(S^2T^*)$. We fix a metric g in M^s, $s > 1$. On the space $H^{s+2}(S^2T^*)$, we can define a cylinder set measure μ^g_N:

$$\mu^g_N(C) \equiv \left(\frac{N}{2\pi}\right)^{\frac{n}{2}} \int_B e^{-\frac{1}{2N}\sum_{k=1}^{n} x_i^2} \prod_{k=1}^{n} dx_i$$

for

$$C \equiv \left\{h \in H^{s+2}(S^2T^*), (< h, h_1 >^g_{s+2}, \ldots, < h, h_n >^g_{s+2}) \in B\right\}$$

$h_1, \ldots, h_n$ are fixed elements in $H^{s+2}(S^2T^*)$.

It follows from the compactness of Λ that the injection $j : H^{s+2}(S^2T^*) \longrightarrow H^s(S^2T^*)$ is a Hilbert Schmidt map (see e.g. [17], [18]). The image measure $j(\mu^g_N)$, which we shall also denote by μ^g_N is thus a σ - additive measure on $H^s(S^2T^*)$ equipped with the Borel σ - algebra ([7],[8],[19],...).

Since the implicit function theorem holds for Hilbert manifolds, we can construct on exponential map around each point g of M^s:

$$\begin{aligned} Exp_g : \quad V \subset T_gM^s &\longrightarrow U_g \subset M^s \\ Y &\longrightarrow Exp_gY \end{aligned} \tag{2.6}$$

where V is an open neighborhood of 0 in T_gM^s and V_g of g in M^s. Following Kuo [8], we can now construct a local measure on M^s.

Proposition 1.1:

Let g be a metric in M^s, B^s be the Borel σ - algebra generated by all open sets of $H^s(S^2T^*)$, then the family of sets:

$$\mathcal{B}^g \equiv \{Exp_g(V \cap B),\ B \in \mathcal{B}^s\} \tag{2.7}$$

is a σ - algebra of subsets of V_g. Moreover, for $N \in \mathbb{N}/\{0\}$ the set function defined on $\mathcal{B}^g$ by:

$$\nu_N^g(D) \equiv \mu_N^g(Exp_g^{-1}(D)) \quad \forall D \in \mathcal{B}^g \tag{2.8}$$

defines a countably additive local measure on M^s around g.

There is no hope of patching up these local measures to define a global measure on M^s since it can be shown using a criterium for the equivalence of Gaussian measures [8],[20] that two measures $\nu_N^{g_1}$ and $\nu_N^{g_2}$ for $g_1, g_2 \in M^s$ are equivalent if and only if the metrics g_1 and g_2 coincide.

This therefore leads to a formulation of a local version of the Polyakov measure which we give in the following paragraph.

2.2 Expectation values with respect to a localized Polyakov measure

On the grounds of the above construction, we define a localized version of the expectation values (2.4) interpreting the formal measure $e^{-\frac{1}{2N}<\tilde{g},\tilde{g}>_g^s} D[\tilde{g}]$ as the local measure ν_N^g and the formal measure $e^{-A(x,g)} D_g[x]$ as a Gaussian measure m^g on the dual of $C^\infty(\Lambda, \mathbb{R}^d)$ w.r.t. to the C^∞ Whitney topology. Let us first introduce the measure m^g on the space of embeddings.

On the space $C^\infty(\Lambda, \mathbb{R}^d)$ of smooth $\mathbb{R}^d$ valued functions Λ, we define the Laplace - Beltrami operator $-\Delta_g^d$ corresponding to a metric g in M^s, $s > 2$ by:

$$-\Delta_g^d x(\eta) \equiv (-\Delta_g x^1(\eta), \ldots, -\Delta_g x^d(\eta)) \tag{2.9}$$

for $x \equiv (x^1, \ldots, x^d)$ in $C^\infty(\Lambda, \mathbb{R}^d)$ and with

$$\Delta_g f(\eta) \equiv \frac{1}{\sqrt{g(\eta)}} \partial_a \sqrt{g(\eta)} g^{ab}(\eta) \partial_b f(\eta) \ \ \forall f \in C^\infty(\Lambda, \mathbb{R}) \tag{2.9'}$$

We shall use the symbol $\Delta_g^{d\prime}$ for the projection of Δ_g^d onto the orthogonal space of the kernel $ker\Delta_g^d$ of Δ_g^d in $L^2(\Lambda, \mathbb{R}^d)$. The space $L^2(\Lambda, \mathbb{R}^d)$ is the closure of $C^\infty(\Lambda, \mathbb{R}^d)$ for the scalar product $(\cdot,\cdot)_g^d$ defined by:

$$(x, y)_g^d \equiv \sum_{\mu=1}^{d} \int_\Lambda \sqrt{g(\eta)} x^\mu(\eta) y^\mu(\eta) d\eta \tag{2.10}$$

(This closure is independent of the choice of the metric g).

The inverse of $\Delta_g^{d\prime}$ will be denoted by $(\Delta_g^{d\prime})^{-1}$. The function Γ defined on $C^\infty(\Lambda, \mathbb{R}^d)$ by:

$$\begin{aligned} \Gamma : \ C^\infty(\Lambda, \mathbb{R}^d) &\longrightarrow \mathbb{R} \\ f &\rightarrow e^{-\frac{1}{2}(f, (\Delta_g^{d\prime})^{-1} f)_g^d} \end{aligned}$$

is positive definite, so that by a theorem by Minlos it defines a countably additive measure m^g with support on the topological dual of $C^\infty(\Lambda, \mathbb{R}^d)'$ of $C^\infty(\Lambda, \mathbb{R}^d)$ with characteristic function given by Γ.It can in fact be shown that this measure is supported by a smaller space, namely the Sobolev space $H^{-1}(\Lambda, \mathbb{R}^d)$ defined by the closure of $C^\infty(\Lambda, \mathbb{R}^d)$ w.r.t the scalar product $\left(x, (\mathbb{1} - \Delta_g)^{-1} x\right)_g^d$ (see e.g [21]). Let us now redefine the expectation values (2.4) taken locally around a fixed metric g in terms of the measures m^g and ν_N^g. (for the sake of simplicity, we shall omit the integration w.r.t. $d\eta_j$.)

For a fixed metric g in M^s, $s > 2$ and for a neighborhood U_g of g in M^s, chosen as in (2.6), we set:

$$\left\langle\left\langle \prod_{j=1}^m S_j \right\rangle\right\rangle_{N,g} \equiv \int_{M^s} \nu_N^g(d\tilde{g}) \mathbb{1}_{U_g}(\tilde{g}) \prod_{j=1}^m \sqrt{\tilde{g}(\eta_j)} \int_{H^{-1}(\Lambda, \mathbb{R}^d)} \prod_{j=1}^m S_j(x, \tilde{g}) e^{i(x, y_{\tilde{g}}^j)_g^d} m^{\tilde{g}}(dx) \tag{2.11}$$

This expectation value gives a localized version of the mean value w.r.t the Polyakov measure of approximated vertex operators described by S_j^N.

We now draw a parallel constrution to the Polyakov procedure in terms of these Gaussian measures.

2.3 The Polyakov procedure in terms of Gaussian measures

The Polyakov procedure for the evaluation of the scattering amplitudes (1.11) requires the following formal operations:

1. A decoupling of the expression (1.11) in the variables x and g so that (1.11) is actually replaced by a family of local decoupled expectations:

$$\int_\Lambda \prod_{j=1}^m d\eta_j \left(\left(\int_{U_g} \prod_{j=1}^m \sqrt{\tilde{g}(\eta_j)} D_g[\tilde{g}] \right) \left(\int_X \prod_{j=1}^m e^{i(x, y_g^j)_g^d} e^{-A(x,g)} D_g[x] \right) \right) \tag{2.12}$$

   where U_g is a local neighborhood of a metric g in M^s and $D_g[\tilde{g}]$ is formal Lebesgue measure on M around the point g.- This formal operation arises in the usual quantization procedure (see e.g. [2],[3],[4],[5],[22],...) when the formal Polyakov measure:

$$e^{-A(x,\tilde{g})} D_{\tilde{g}}[x] D[\tilde{g}]$$

   around a point g is replaced by the expression:

$$\left(e^{-A(x,g)} D_g[x]\right) D_g[\tilde{g}]$$

   where the measure on the space of metrics and on the space of embeddings are decoupled.

2. A transformation of the measure $D[g]$ through a parametrization of the space of smooth Riemannian metrics in terms of smooth diffeomorphisms, smooth Weyl transformations and the Teichmüller space on Λ.

These two steps can be given a precise meaning for the approximated expectation values $<< \prod_{j=1}^m S_j >>_{N,g}$ in the simple case of tachyons $S_j = 1 \ \forall j = 1, \ldots, m$.

The decoupling in the variables x and g follows from a mean value theorem combined with the continuity of the map:

$$M^s \longrightarrow \mathbb{R}$$
$$g \to \int_{H^{-1}(\Lambda, \mathbb{R}^d)} e^{i(x, y_g^j)_g^d} m^g(dx)$$

The expectation (2.11) then splits up into the product of two expectations since we have:

$$\left\langle\left\langle \prod_{j=1}^m S_j \right\rangle\right\rangle_{N,g} = \left[\int_{H^{-1}(\Lambda, \mathbb{R}^d)} \prod_{j=1}^m e^{i(x, y_g^j)_g^d} m^g(dx)\right] \left[\int_{M^s} \nu_N^g(d\tilde{g}) 1\!\!1_{U_g}(\tilde{g}) \prod_{j=1}^m \sqrt{\tilde{g}(\eta_j)}\right] + \mathcal{E}(U_g)$$

where $\mathcal{E}(U_g)$ tends to zero when the diameter of U_g (taken w.r.t. the scalar product $< \cdot, \cdot >_s^g$).

The transformation of the measure ν_N^g on the space of metrics results from a continuous parametrization of the space M^s in terms of the space of diffeomorphisms homotopic to identity of Sobolev class $s+1$, the space of Weyl transformations of Sobolev class s and the Teichmüller space $\mathcal{T}^s$ described as the space of non equivalent complex structures on Λ of Sobolev class s. This parametrization is only continuous since the action of the group of diffeomorphisms of Sobolev class $s+1$ on M^s is only continuous. It becomes smooth when restricted to the space M of smooth Riemannian metrics (see e.g. [9]).

This mathematical construction in the frame work of Sobolev Riemannian metrics motivates the formal procedure we briefly describe in the following chapter in the case of smooth Riemannian metrics by which the quantization of closed bosonic strings is put into relation with the quantization of the Liouville model when $d \leq 12$.

3. An interpretation of the Polyakov model in terms of the Liouville measure

In the first chapter, we pointed out that the conformal invariance which holds at the classical level breaks up at the quantized level.This gives rise to a formal measure on the Weyl group in the reduction of the "Polyakov measure". This term due to the conformal anomaly formally disappears in the critical dimension ($d = 26$)(e.g.[14] ,[15],[22],[26],[27]).

The object of this chapter is an interpretation for non critical dimension, here for space - time dimension $d \leq 12$, of this measure on the Weyl group in terms of a well defined family of Liouville measures indexed by the Teichmüller parameters. We shall summarize here results of [1] in which this correspondance between the quantization of the closed bosonic strings and the quantization of the classical Liouville model was establisched in full details.

Following the usual Polyakov procedure, one first parametrizes the space of Riemannian metrics in terms of the Weyl group, the group of diffeomorphisms homotopic to identity and the Teichmüller space. For this we extend in the first paragraph the result stated in the preceeding chapter concerning the existence of a continuous parametrization of the space M^s to the case $s = +\infty$, for which this parametrization becomes smooth.

3.1 A smooth parametrization of the space of smooth Riemannian metrics

We recall that the genus p of Λ is strictly larger than 1 and introduce the space M_{-1} of smooth Riemannian metrics on Λ with scalar curvature equal -1. M_{-1} has the structure of a Fréchet manifold (see e.g. [23]) and the Fréchet Lie group D_0 of smooth diffeomorphisms homotopic to identity acts smoothly on M_{-1}. The quotient space M_{-1}/D_0 interits the Fréchet structure of M_{-1}. The existence of a smooth parametrization of M is essentially a consequence of the existence of a global smooth section of the bundle M_{-1} over the Teichmüller space $\mathcal{T}$ of Λ consisting of non equivalent smooth complex structures on Λ with structure group D_0. Such a global smooth section was explicitly constructed in [10] and further investigated in [23]. We shall denote it by $S : \mathcal{T} \to M_{-1}$. This yields a natural description of M in terms of $D_0, \mathcal{T}$ and P (see (1.8) using the diffeomorphism between M_{-1} and the quotient Fréchet manifold M/D_0.

Theorem 3.1:

For a metric g in M, there exists a unique t in $\mathcal{T}$, a unique element f in D_0, and a unique element φ in $C^\infty(\Lambda, I\!R)$ such that:

$$g = f^* e^\varphi g_t \qquad \text{where } g_t \equiv S(t)$$

Moreover the map:

$$\begin{aligned} T : C^\infty(\Lambda, I\!R) \times D_0 \times \mathcal{T} &\longrightarrow M \\ (\varphi, f, t) &\rightarrow g \equiv f^* e^\varphi g_t \end{aligned}$$

is smooth.

The map T being smooth,induces on each tangent space $T_g M$ to M at point g a differential map dT^g which can be computed explicitely in terms of elliptic operators acting on $C^\infty(\Lambda, I\!R)$ and $C^\infty(T\Lambda)$, the space of smooth sections of the tangent bundle to Λ denoted by $T\Lambda$ (see e.g. [1],[22],[24]).

On the grounds of the discussion in chapter 2, identifying locally the formal local measure $D[g]$ on M occuring in (1.10) with a measure $D_g[h]$ on the tangent space $T_g M$ to M at point g, taking into account the decoupling of the measure $D_g[x]$ on the space of embeddings from the measure $D[g]$ on the space of Riemannian metrics, the description of scattering amplitudes for tachyons boils down to the study of a formal measure of the form:

$$\left(\int_{C^\infty(\Lambda,\mathbb{R}^d)} \prod_{j=1}^{m} e^{i(x,y_g^j)_g^d} e^{-\frac{1}{2}(x,\Delta_g^d x)_g^d} D_g[x]\right) D_g[h] \tag{3.1}$$

Here, g is a fixed metric in M.

The transformation of the expression (3.1) through a parametrization of M – the measure $D_g[h]$ being transformed through the differential operator $d\mathcal{T}^g$ – leads to a product of formal measures on the product of the tangent spaces to the group P of smooth Weyl transformations, to the group D_0 of smooth diffeomorphisms homotopic to identity and to the Teichmüller space $\mathcal{T}$.

We now introduce the Liouville measure in order to express the amplitudes (3.1) after transformation under the reparametrization of M.

3.2 The Liouville measure

The quantization of the Liouville model in 2 dimensions which is classically described by the Lagrangian density:

$$\mathcal{L}(\varphi(t,x)) \equiv \frac{1}{2}\left(\frac{\partial\varphi}{\partial t}\right)^2 - \frac{1}{2}\left(\frac{\partial\varphi}{\partial x}\right)^2 - \lambda e^{\alpha\varphi}, \ \ (t,x)\in\mathbb{R}^2, \ \lambda>0, \ \alpha\in\mathbb{R}$$

has been investigated by S. Albeverio and R. Hoegh - Krohn in [25], where the authors show that for $\alpha^2 < 4\pi$, the measure describing the functional quantization of this model is a well defined probability measure.

The methods used in [25] carry on to the case of a two dimensional real Riemannian manifold, since the upper bound of α is dictated by the singularities of the Green function and hence only depends on the local properties of the two dimensional real manifold.

More precisely, let $\mu_{0,g}$ denote the free field measure on $C^\infty(\Lambda,\mathbb{R})'$, the topological dual of $C^\infty(\Lambda,\mathbb{R})$ equipped with the C^∞ Whitney topology, determined by the characteristic function:

$$\hat{\mu}_{0,g}(f) \equiv e^{-\frac{1}{2}\int_\Lambda \sqrt{g(\eta)} f(\eta)(-\Delta_g'^{-1} f)(\eta)d\eta}$$

where g is a fixed metric in m, then, for $\alpha^2 < 4\pi$ the expression:

$$d\mu_g(\xi) \equiv \left(\int e^{-\lambda U}(\Lambda,g)^{(\xi)} d\mu_{0,g}(\xi)\right)^{-1} e^{-\lambda U_{(\Lambda,g)}(\xi)} d\mu_{0,g}(\xi)$$

is a well defined probability measure. Here, $U_{(\Lambda,g)}(\xi) \equiv: e^{\alpha\xi} : g(\chi_\Lambda)$, whereby $: \ :_g$ denotes the Wick product relative to the Riemannian structure given by g on Λ and χ_Λ the characteristic function of Λ.

3.3 The correspondance between the quantized Liouville model and the quantization of bosonic strings

Introducing a cosmological term (already mentioned in paragraph 1.2) $e^{-\mu^2 \int_\Lambda \sqrt{g(\eta)}d\eta}, \mu^2 > 0$ in the expression (3.1) (let us remark that this term is invariant under the action of the diffeomorphism group), and setting $\alpha^2 = (\frac{26-d}{48\Pi})^{-1}$, where d is the dimension of space - time, one can interpret the scattering amplitudes for tachyons as follows:

$$\int_T H(t)dt \int_{\Psi\equiv\frac{\varphi}{\alpha}} F(\Psi, g_t)d\mu_{g_t}(\Psi) \tag{3.2}$$

$F(\Psi, g_t)$ is a term that stands for the contributions of the vertex operators for tachyons, $H(t)dt$ is the measure on the Teichmüller space corresponding to the reduction in critical dimension ($d = 26$), g_t is the reference metric given by Theorem 3.1 for a fixed Teichmüller parameter t in T. The parameter φ arises from the parametrization of M given by this same theorem.

The measure $H(t)dt$ on the Teichmüller space obtained by reduction of the Polyakov measure in critical dimension has been investigated in the literature on strings in the last few years (see e. g. [26],[27],...).

The next step towards an interpretation of the Polyakov model w.r.t. the Liouville model in non critical dimensions is a study of the Teichmüller dependance of the Liouville measure μ_{g_t}, $t \in T$ in order to give a meaning to the formal expression $[H(t)d\mu_{g_t}(\Psi)]\, dt$. This will be the object of further articles by the coauthors of [1].

REFERENCES

[1] S. Albeverio, R. Hoegh - Krohn, S. Paycha, S. Scarlatti, Path space measure for the Liouville quantum field theory and the quantization of relativistic strings, Phys. Letters B 174,(1986),p. 81-85
S. Albeverio, R. Hoegh - Krohn, S. Paycha, S. Scarlatti, paper in preparation.

[2] O. Alvarez, Theory of strings with boundaries: fluctuations, topology and quantum geometry, Nucl. Phys. B 216 (1983), p. 125.

[3] B. Durhuus, Quantum theory of strings, Nordita preprint, Lectures given at NBI / Nordita, autumn 1982 (non published)

[4] J. Polchinski, Evaluation of the one loop string path integral, C.M.P. 104 (1986), p. 37-47

[5] A. Polyakov, Quantum geometry of bosonic strings, Phys Lett. 103 B, (1981) p. 207.

[6] K. D. Elworthy, Measures on infinite dimensional manifolds. Functional integration and its applications, (A. M. Arthurs, Ed.), Oxford University Press (1975), p. 61-67.

[7] L. Gross, Measurable functions on Hilbert spaces, T.A.M.S. 105, (1962), p. 372-390
L. Gross, Abstract Wiener spaces, Proc. Fifth Berkeley Symp. in Math. Stat. and Probability (1965/66), p. 31-42.

[8] H. H. Kuo, Gaussian measures in Banach spaces, Lect. Notes in Math. 463, Springer-Verlag.
H. H. Kuo, Integration theory on infinite dimensional manifolds, F.A.M.S., Vol. 159, (1971), p. 57-78.

[9] D. Ebin, J. Marsden, Groups of diffeomorphisms and the motion of an incompressible fluid, Ann. Math. 92 (1970), p. 102-163.

[10] C. J. Earle, J. Eells, A fibre bundle description of Teichmüller theory, J. Diff. Geom. 3, (1969),p. 19-43.

[11] S. Paycha, Probability measures on infinite dimensional manifolds and Polyakov strings, Thesis (1988)

[12] Y. Nambu in Symmetries and Quark Models, ed. R. Chand, Gordon and Breach, New York 1970.
T. Goto, Progr. Theor. Phys. 46 (1971), p. 1560

[13] L. Brink, P. D. Vecchia, P. Howe, A locally supersymmetric reparametrization invariant action for the spinning string, Phys. Lett. B 65, (1976), p. 471

[14] M. G. Green, T. H. Schwartz, E. Witten, Superstring theory, Vol. 1, Cambridge Monographs on mathematical physics (1987).

[15] S. Weinberg, Covariant path integral approach to string theory, lectures at 3rd Jerusalem winter school of theoretical physics (1987), Preprint UTT 6 - 17 - 87.

[16] S. Hamilton, Nash Moser inverse function theorem, B.A.M.S. Vol.7 (1982), p. 65-222.

[17] J. Eells, Integration on Banach Manifolds, Proceedings of the 13-th Biennal Seminar of the canadian mathematical congress, Halifax, (1971), p. 41-49.

[18] K. Maurin, Methods of Hilbert spaces, Mon. Mat. Warsaw (1967).

[19] P. Baxendale, Gaussian measures on function spaces, Amer. J. Math. Vol.98, No. 4 (1976), p. 891-952.

[20] A. V. Skorohod, Integration in Hilbert spaces, Ergebnisse der Mathematik und ihrer Randgebiete, Springer - Verlag (1974).

[21] Y. Yamaski, Measures on infinite dimensional spaces, Series in pure mathematics Vol.5, World Scientific (1984).

[22] E. Hoker, D. Phong, Multiloop amplitudes for the bosonic Polyakov string, Nucl. Phys. B 269, p. 205 (1986).

[23] A. E. Fischer, A. S. Tromba, On a purely "Riemannian" proof of the structure and dimension of the unramified moduli space on a compact Riemann surface, Math. Anal. 267, p. 311-345 (1984).

[24] G. Moore, P. Nelson, Measure for Moduli, Nucl. Phys. B 266, (1986), p. 58.

[25] S. Albeverio, R. Hoegh - Krohn, The Wightman axioms and the mass gap for strong interactions of exponential type in two dimensional space - time, J. Funct. Anal. 16, p. 39-82.

[26] L. Alvarez - Gaumé, P. Nelson, Riemann surfaces and string theory, Supersymmetry, Supergravity and Superstrings 86 (World Scientific, 1986).

[27] D. S. Smit, String theory and algebraic geometry of moduli spaces, Com. Mat. Phys. 114 (1988), p. 645-685.

A REACTION-DIFFUSION MODEL FOR MODERATELY INTERACTING PARTICLES

H. ROST
Institut für Angewandte Mathematik
Universität Heidelberg
Im Neuenheimer Feld 294
D-6900 Heidelberg 1

ABSTRACT. A certain class of models of n independently migrating particles with pair interaction is analyzed; it is shown that for large n in a suitable limit a certain non-linear reaction-diffusion equation appears, independent of the particular scaling sequence. That way the concept of what we call "moderate interaction" is clarified.

0. PRELIMINARY REMARK

We report here on results of a joint work with G. Nappo and E. Orlandi; more details may be found in the paper [3] .

1. VARIOUS TYPES OF REACTION-DIFFUSION MODELS

Perhaps the simplest reaction-diffusion model, which in a suitable limit will lead to a non-linear equation is of the following kind: particles move independently in R^d according to a Brownian law; the only reaction between them is a binary one: whenever two particles come close to each other one of them disappears and will never come back. If one wants to make precise what "close to each other" means, there are two choices. One consists in interpreting the position of a particle as the center of ball with radius r and in removing at random one of two balls at the moment where they touch each other. The other possibility is to introduce a new random mechanism describing the death (removal) of the ith particle: it dies with a rate

$$r_i(t) = \frac{1}{n} \sum_{j \neq i} q(X_i(t) - X_j(t)) \quad . \tag{1}$$

Here X_i and X_j denote the positions of particle i and j , in the number present at time t = 0, and q is some positive function on R^d describing the reaction rate of two molecules as a function of their relative position; the summation is extended over all indices j such that particle j is still alive at time t .

S. Albeverio et al. (eds.), Stochastics, Algebra and Analysis in Classical and Quantum Dynamics, 205–209.

To state it more formally: we assume that X_i, $1 \leq i \leq n$ are independent Brownian motions with identical initial law $f_o(x)dx$; since neither f_o nor the Brownian law are changed with n, we may (in order to "couple" the processes for different n in an easy way) assume that X_i, $1 \leq i \leq n$, is the beginning of an infinite sequence of i.i.d. processes X_i, $i \geq 1$. We define by $\xi_{i,n}(t)$ the indicator of the event that particle i is alive at time t. Then one is interested in the limiting behavior, as $n \to \infty$, of the empirical measures

$$M_n(t) = \frac{1}{n} \sum_{i=1}^{n} \delta_{X_i(t)} \cdot \xi_{i,n}(t) . \tag{2}$$

In order to get a non-degenerate limit, in either choice of the reaction mechanism the parameters p or q have to depend in a reasonable way on n. For the first case, that of instantaneous death, the question has been solved by Sznitman [5]. He shows that the correct dependence is

$$\begin{aligned} d = 2 \quad &: \quad p_n = \exp(-c_1 n) \\ d \geq 3 \quad &: \quad p_n = c_2\, n^{-1/d-2} \end{aligned} \tag{3}$$

(For $d = 1$, a non-trivial limit does not exist, obviously, in that model.) Under (3), the sequence $M_n(t)$ has a deterministic limit $u(t,x)dx$, which solves the equation

$$\begin{aligned} \frac{\partial u}{\partial t} &= \frac{1}{2} \Delta u - c_3 u^2 \\ u(0,\cdot) &= f_o \end{aligned} \tag{4}$$

In the second case, that of "continuously accumulated death risk", one may fix some positive function Q and set

$$q_n(x) = a_n^d \cdot Q(a_n x) . \tag{5}$$

Then again two principally different ways of chosing (a_n) exist: one may put $a_n = p_n^{-1}$ with p_n given by (3); then one is close to [5] and expects a limit u of M_n, obeying an equation (4) with possibly another constant c_4 instead of c_3, but such that c_4 depends in a rather complicated way on the geometry of the function Q. The other way is what we are going to study here, namely to let (a_n) grow slower than p_n^{-1}. We call such a choice of a_n, resp. q_n, the case of moderate interaction. It turns out that one then gets again a limit u, given by

$$\frac{\partial u}{\partial t} = \frac{1}{2} \Delta u - cu^2 \tag{6}$$

$$u(0,\cdot) = f_o \quad ;$$

but now

$$c = \int Q(x)dx \tag{7}$$

holds, regardless of the special sequence (a_n) as long as it belongs to the régime of moderate interaction.

2. RESULTS

The precise formulation of our result is as follows. We introduce the assumptions:

(A.1) Q is integrable, positive, and in H^{-1}, i.e. its Fourier transform $\hat{Q}$ satisfies

$$\int |\hat{Q}(\omega)|^2 (1 + |\omega|^2)^{-1} d\omega < \infty \quad .$$

(A.2) The sequence (a_n) satisfies

if $d = 1$, $a_n \to \infty$ (no other restriction)

if $d = 2$, $\log a_n = o(n)$

if $d \geq 3$, $a_n = o(n^{1/d-2})$.

(A.3) The initial density f_o is bounded in supremum norm.

Theorem. Under (A.1) to (A.3), the empirical measures $M_n(t)$ converge weakly in probability to $u(t,x)dx$ given by (6) and (7).

We show a slightly sharper result, namely "propagation of chaos", i.e. asymptotic independence of the survival of different particles in the following sense:

Consider the sequence of i.i.d. processes X_i introduced above; define processes ξ_i, $i \geq 1$, equal to 1 up to some random time and equal to 0 afterwards, in such a way that the intensity of a jump of ξ_i from 1 to 0 at time t, given all $X_i(s)$, $s \leq t$, is equal to $C \cdot u(t,X_i(t))$, u and C as in (6) and (7). Then for any fixed finite set F of indices one has convergence of $(X_i,\xi_{i,n})_{i \in F}$ to $(X_i,\xi_i)_{i \in F}$, in distribution with respect to the Skorohod metric in $\mathcal{D}([0,\infty),(R \times \{0,1\}))^F$. Obviously the limit law is a product measure, i.e. "chaotic".

3. PROOF: AN INCLUSION-EXCLUSION PRINCIPLE

The main idea of the proof is to analyze the $\xi_{i,n}$ and the corresponding "cumulative risk processes" $R_{i,n}$, where

$$R_{i,n}(t) = \int_0^t ds \cdot \frac{1}{n} \sum_{\substack{j=1 \\ j\neq i}}^{n} q_n(X_i(s) - X_j(s)) \cdot \xi_{j,n}(s) . \tag{8}$$

One sees that they can be approximated from above and below in an alternating way by processes $\xi^k_{i,n}$ and $R^k_{i,n}$, in the following sense. Let σ_i, $i \geq 1$, be an i.i.d. sequence of exponential variables with mean 1; define on the given probability space recursively for $k \geq 0$

$$R^0_{i,n}(t) = 0 \tag{9}$$

$$\xi^k_{i,n}(t) = 1_{\{R^k_{i,n}(t) < \sigma_i\}}$$

$$R^{k+1}_{i,n}(t) = \int_0^t ds \cdot \frac{1}{n} \sum_{\substack{j=1 \\ j\neq i}}^{n} q_n(X_i(s) - X_j(s)) \cdot \xi^k_{j,n}(s) .$$

If we repeat that procedure on the level of partial differential equations instead of processes we come to functions u^k, which over- and underestimate u in an alternating way and are defined for $k \geq 0$ by

$$u^0(t,x) = 0 \tag{10}$$

u^{k+1} is solution to the linear equation

$$\frac{\partial v}{\partial t} = \frac{1}{2} \Delta v - c \cdot u^k \cdot v \ ; \ v(0,\cdot) = f_0$$

Set finally

$$R^k_i(t) = c \cdot \int_0^t u^k(s, X_i(s))ds \tag{11}$$

$$R_i(t) = c \cdot \int_0^t u\ (s, X_i(s))ds$$

Since one can sandwich u between two consecutive u^k with arbitrary precision if k is large, to prove that $R_{i,n}$ converges to R_i (and hence $\xi_{i,n}$ to ξ_i) is reduced to the proof of convergence of $R^k_{i,n}$ to R^k_i for each fixed k .

4. CUMULATIVE RISK AS THE ESSENTIAL NOTION

The method of considering the cumulative risk function $R_{i,n}$ rather than its time derivative, the hazard rate (as for example in [1]), allows us to get results that are close to being sharp what concerns the assumptions on (a_n), as comparison of (A.2) and (3) shows. It seems to us that the $R_{i,n}$ are the natural objects to look at, at least in the régime of moderate interaction. One may even speculate that a good definition of what one should call in this context "moderate interaction" is to characterize it by the property that in the limit $n \to \infty$ the risk $R_{i,n}$ becomes a deterministic function of the process X_i only.

5. HISTORICAL REMARKS

Lang and Nguyen [2] drew our attention to the by now classical work of Smoluchowski [4] , where reaction-diffusion models already appear, with a reaction mechanism, however, corresponding to the model studied in [5] .

6. REFERENCES

[1] Dittrich P.
A stochastic model of a chemical reaction with diffusion. Prob.Theory rel.fields 79, 115-128 (1983).

[2] Lang R. and Nguyen XX.
Smoluchowski's theory of coagulations in colloid holds rigorously in the Boltzmann-Grad limit. Z.Wahrscheinlichkeitstheorie verw.Geb., 54, 227-280 (1980).

[3] Nappo G., Orlandi E. and Rost H.
A reaction-diffusion model for moderately interacting particles (to appear in: J.Stat.Physics (1989)).

[4] Smoluchowski, M. von
Versuch einer mathematischen Theorie der Koagulations-kinetik kolloider Lösungen. Z.Phys.Chemie 92, 129-168 (1917).

[5] Sznitman A.S.
Propagation of chaos for a system of annihilating Brownian spheres. Comm.Pure Appl.Math. 40, 663-690 (1987).

Nondegeneracy in the Perturbation Theory of Integrable Dynamical Systems

Helmut Rüssmann
Fachbereich Mathematik
Johannes Gutenberg-Universität Mainz
D-6500 Mainz
Bundesrepublik Deutschland

ABSTRACT. The most general nondegeneracy condition for the existence of invariant tori in nearly integrable and analytic Hamiltonian systems is formulated.

1. The Problem

The prototype of a perturbed integrable dynamical system is a Hamiltonian system

$$\begin{aligned} \frac{dx}{dt} &= \frac{\partial H}{\partial y} = \left(\frac{\partial H}{\partial y_1}, \dots, \frac{\partial H}{\partial y_n}\right) \\ \frac{dy}{dt} &= -\frac{\partial H}{\partial x} = -\left(\frac{\partial H}{\partial x_1}, \dots, \frac{\partial H}{\partial x_n}\right) \end{aligned} \tag{1}$$

where

$$x = (x_1, \dots, x_n) \in T^n = \mathbb{R}^n / 2\pi\mathbb{Z}^n$$

is the vector of angular variables and

$$y = (y_1, \dots, y_n) \in B \subseteq \mathbb{R}^n$$

is the vector of action variables varying in some open set B of $\mathbb{R}^n$. The Hamiltonian

$$H : T^n \times B \longrightarrow \mathbb{R}$$

is supposed to be real analytic and of the form

S. Albeverio et al. (eds.), Stochastics, Algebra and Analysis in Classical and Quantum Dynamics, 211–223.

$$H(x,y) = H_0(y) + H_1(x,y) \tag{2}$$

where

$$|H_1(x,y)| \leq \varepsilon \qquad \text{for all } x \in T^n, \; y \in B \tag{3}$$

with $\varepsilon > 0$ sufficiently small.
The unperturbed system ($\varepsilon = 0$)

$$\frac{dx}{dt} = \omega(y) := \frac{\partial H_0}{\partial y}(y) \quad , \quad \frac{dy}{dt} = 0$$

has nothing but quasiperiodic phase trajectories

$$x = t\omega(b) + \text{const.} \tag{4}$$

lying on invariant tori

$$y = b \in B \, . \tag{5}$$

In the case that the frequency vector $\omega(b) = (\omega_1(b),\dots,\omega_n(b))$ is nonresonant, that is

$$< k,\omega(b)> = k_1\omega_1(b)+\dots+k_n\omega_n(b) \neq 0 \quad \text{for all}$$

$$k = (k_1,\dots,k_n) \in \mathbb{N}_0^n \setminus \{0\} \; , \; \mathbb{N}_0 = \{0,1,2,\dots\} \, ,$$

the quasiperiodic phase trajectories (4) fill the invariant torus $T^n x\{b\}$, on which they lie, densly.
The question is under which conditions these invariant tori survive in the perturbed system ($0<\varepsilon<<1$). The survival can roughly be described as follows:

The torus $T^n x\{b\}$ is slightly deformed,

$$y = b + U(x,b) \tag{6}$$

where

$$U: T^n xB \longrightarrow \mathbb{R}^n$$

is continuous, and we have

$$\sup_{x\in T^n} |U(x,b)| \longrightarrow 0 \quad \text{for } \varepsilon \longrightarrow 0 \text{ and } b \in B. \tag{7}$$

Moreover $U=(U_1,\dots,U_n)$ has partial derivatives with respect to $x_1,\dots,x_n$, and we have

$$\frac{\partial U_j}{\partial x_l} = \frac{\partial U_l}{\partial x_j} \; .$$

The frequency vector is slightly deformed, but nonresonant, such that the phase trajectories

$$x = t\bar{\omega}(b) + \text{const.} \; , \; t \in \mathbb{R} \tag{8}$$

fill (6) also densly, and $\bar{\omega}$: B —> $\mathbb{R}$ is supposed to be at least of class C^1 and to satisfy

$$|\omega(b) - \bar{\omega}(b)| \longrightarrow 0 \quad \text{for} \quad \varepsilon \longrightarrow 0 \tag{9}$$

The well known theorem of A.N. Kolmogorov [**4**, p. 183] states that a sufficient condition for the survival of most of the invariant tori (5) in the form (6), (7) is

$$\det \frac{\partial\omega}{\partial y}(b) = \det \frac{\partial^2 H_0}{\partial y_j \partial y_l}(b) \neq 0 \text{ for all } b \in B. \tag{10}$$

As Moser [**8**,p.44] has shown this theorem is true even with $\bar{\omega} = \omega$. The goal of this paper is to formulate infinitely many further sufficient conditions which all together lead to a very natural nondegeneracy condition for ω resp. H_0 guaranteeing the existence of invariant tori of the perturbed system (1),(2),(3), $0<\varepsilon<<1$.

2. A General Nondegeneracy Condition

In the following definition the frequency vector ω is not necessarily the gradient of a Hamiltonian H_0 as above.

Definition 1: Let B be an open and connected subset of $\mathbb{R}^n$ and ω:B —> $\mathbb{R}^n$ be a real analytic vector function. We call ω nondegenerate if the range $\omega(B)$ of ω does not lie in an (n-1)-dimensional linear subspace of $\mathbb{R}^n$. If $\omega(B)$ belongs to such a subspace ω is called degenerate.

Remark: In the case n=1 a real analytic function ω:B —> $\mathbb{R}$ with an open interval $B \subseteq \mathbb{R}$ is nondegenerate if and only if it does not vanish identically in B. Now from function theory it is well known that if ω vanishes in B then at each point $b \in B$ all coefficients of the Taylor expansion of ω in b vanish and conversly, if all coefficients of the Taylor expansion of ω vanish in some point $b \in B$ then ω vanishes identically in B. So for nondegeneracy of ω in the case n=1 it is sufficient that the Taylor expansion of ω in some point $b \in B$ contains at least one non-vanishing coefficient, and necessary that for each point $b \in B$ the Taylor expansion of ω in b contains at least one non-vanishing coefficient.

The corresponding facts for arbitrary n are formulated in the following

Lemma 1: Let $B \subseteq \mathbb{R}^n$ be open and connected. Then for a real analytic function $\omega : B \longrightarrow \mathbb{R}^n$ to be nondegenerate it is sufficient that the Taylor expansion

$$\omega(y) = \sum_{l \in \mathbb{N}_0^n} \frac{(y-b)^l}{l!} \omega^{(l)}(b) \tag{11}$$

of ω in some point $b \in B$ contains n linearly independent coefficients

$$\omega^{(l)}(b) \in \mathbb{R}^n , \quad l = l^{(1)} , \ldots, l^{(n)} \in \mathbb{N}_0^n , \tag{12}$$

and it is necessary that the Taylor expansion (11) contains n linearly independent coefficients (12) at each point $b \in B$ where the choice of the indices $l^{(\alpha)} = l^{(\alpha)}(b)$, $\alpha = 1, \ldots, n$ may depend on $b \in B$.

Proof: We prove the corresponding statement for degeneracy. We fix b and assume that the series (11) does not contain n linearly independent coefficients (12). Then the linear space V spanned by all Taylor coefficients $\omega^{(l)}(b)$, $l \in \mathbb{N}_0^n$ has dimension $\leq n-1$. We consider $\mathbb{R}^n$ as an euclidean space provided with the usual inner product $\langle .,. \rangle$ and choose an orthonormal base $\{e_1, \ldots, e_n\}$ such that V is contained in the linear subspace spanned by $e_1, \ldots, e_{n-1}$. Then we can write

$$\omega^{(l)}(b) = \sum_{j=1}^{n-1} a_{jl} e_j \quad , \quad l \quad \mathbb{N}_0^n \tag{13}$$

with uniquely determined $a_{jl} \in \mathbb{R}$. Inserting (13) in (11) we get

$$\omega(y) = \sum_{j=1}^{n-1} \omega_j(y) e_j \tag{14}$$

with

$$\omega_j(y) = \sum_{l \in \mathbb{N}_0^n} \frac{(y-b)^l}{l!} a_{jl} \quad , \quad j = 1, \ldots, n-1 .$$

From (13) we obtain the estimate

$$|\omega^{(1)}(b)| = \left(\sum_{d=1}^{n-1} a_{j1}^2\right)^{\frac{1}{2}} \geq |a_{j1}|$$

which makes clear that each series $\omega_j(y)$ is absolutely convergent in every polydisc

$$|y-b| = \max_{1\leq j\leq n} |y_j-b_j| \leq r$$

in which the series (11) converges absolutely. So $\omega_1,\ldots,\omega_{n-1}$ are analytic in a certain open neighbourhood W of b.

Now ω is analytic in B. Therefore the functions

$$y \longrightarrow \omega_j(y) = <\omega(y), e_j> , \; j = 1,\ldots,n$$

are analytic in B. Comparing

$$\omega(y) = \sum_{j=1}^{n} \omega_j(y)e_j$$

with (14) we get $\omega_n(y) = 0$ in the neighbourhood W of b. Since B is connected we have $\omega_n(y) = 0$ for all $y \in B$. As a consequence $\omega(B)$ lies in an (n-1)-dimensional linear subspace of $\mathbb{R}^n$. Conversly, if this is the case we choose an orthonormal base $\{e_1,\ldots,e_{n-1}\}$ of this subspace. So we obtain (14) with coefficients

$$y \longrightarrow \omega_j(y) = <\omega(y), e_j> , \; j = 1,\ldots,n-1$$

analytic in B. Now we differentiate (14) at an arbitrary point $b \in B$ in order to obtain (13) with

$$a_{j1} = \omega_j^{(1)}(b) .$$

But (13) means that all Taylor coefficients at b belong to an (n-1)-dimensional linear subspace of $\mathbb{R}^n$. So n linearly independent coefficients (12) cannot be found. The lemma is proved.

<u>Definition 2:</u> Let B be an open connected subset of $\mathbb{R}^n$ and $H_0 : B \longrightarrow \mathbb{R}$ be a real analytic function. We call H_0 nondegenerate if the gradient

$$y \longrightarrow \omega(y) = \frac{\partial H_0}{\partial y}(y) : B \longrightarrow \mathbb{R}^n$$

of H_0 is nondegenerate: Otherwise H_0 is called degenerate.

Lemma 2: Let $B \subseteq \mathbb{R}^n$ be open and connected. A real analytic function $H_0 : B \longrightarrow \mathbb{R}$ is nondegenerate if and only if it does not depend - up to a linear change of variables - on less than n variables.

Proof: H_0 is degenerate if and only if the range of $\omega = \frac{\partial H_0}{\partial y}$ lies in an (n-1)-dimensional subspace V of $\mathbb{R}^n$. We assume that this is the case and choose an orthonormal base $\{e_1,\dots,e_n\}$ of $\mathbb{R}^n$ such that V lies in the span of $e_1,\dots,e_{n-1}$. Then we have

$$\langle \omega(y), e_n \rangle = 0 \quad \text{for all } y \in B, \tag{15}$$

hence, putting

$$y = zC = z_1 e_1 + \dots + z_n e_n , \tag{16}$$

$$\frac{\partial H_0(zC)}{\partial z_n} = \langle \frac{\partial H_0}{\partial y}(zC),\ e_n \rangle = 0 \quad \text{for all } z \in BC^{-1}. \tag{17}$$

So the function

$$z = (z_1,\dots,z_n) \longrightarrow H_0(zC)$$

does not depend on z_n in its domain of definition BC^{-1}. Conversly if we have (17) for a linear transformation (16) where only

$$\det(e_1,\dots,e_n) \neq 0$$

is required, we get (15), and consequently $\omega(B)$ lies in an (n-1)-dimensional subspace of $\mathbb{R}^n$. The lemma is proved.

Examples: We put $B = \mathbb{R}^n$ and define H_0 by

1. $H_0(y) = y_1 y_1 + y_2 y_1^2 + y_3 y_1^3 + \dots + y_n y_1^n$ is nondegenerate by lemma 1 because we have

$$\omega(y) = \frac{\partial H_0}{\partial y}(y) = (2y_1, y_1^2, \dots, y_1^n) ,$$

and so

$$\det\left(\frac{\partial\omega}{\partial y_1}, \frac{\partial^2\omega}{\partial y_1^2}, \dots, \frac{\partial^n\omega}{\partial y_1^n} \right)$$

$$= \begin{vmatrix} 2 & \cdot & \cdot & \cdot & \cdot & \cdot \\ & 2! & \cdot & \cdot & \cdot & \cdot \\ & & 3! & \cdot & \cdot & \cdot \\ & 0 & & \cdot & & \\ & & & & \cdot & \\ & & & & & n! \end{vmatrix} = 2(2!)\dots(n!) \neq 0.$$

2. $H_0(y) = (y_1+\dots+y_n)y_1^2$. Here H_0 is degenerate for $n > 2$ by lemma 2. For $n=2$ H_0 is nondegenerate by lemma 1 because, for example,

$$\det\left(\omega, \frac{\partial\omega}{\partial y_2}\right)(1,1) = \begin{vmatrix} 5 & 2 \\ 1 & 0 \end{vmatrix} = -2 .$$

3. Formulation of the Existence Theorem

We have the following

Theorem: Let B be an open connected subset of $\mathbb{R}^n$ and let the unperturbed Hamiltonian $H_0 : B \longrightarrow \mathbb{R}$ of system (1),(2),(3) be

nondegenerate. Then most of the invariant tori $T^n \times \{b\}$, $b \in B$ of the unperturbed system ($\varepsilon = 0$) survive in the form (6),(7). These surviving tori form a majority in the sense that given any open set K with compact closure $K \subseteq B$ then the Lebesgue measure of the set

$$K_\varepsilon = \left\{ b \in K \;\middle|\; \begin{array}{l} \text{there is an invariant torus (6) of} \\ \text{(1),(2),(3) with (7)} \end{array} \right\}$$

tends to the full measure of K:

$$\mu(K_\varepsilon) \nearrow \mu(K) \qquad \text{for} \qquad \varepsilon \longrightarrow 0.$$

Moreover the frequency vector ω of the flow (8) on the invariant torus (6) , $b \in K_\varepsilon$ satisfies the relations (9) and

$$|\langle k, \bar{\omega}(b) \rangle| \geq c(\varepsilon) e^{-|k|^{\frac{1}{2}}} \qquad \text{for all} \quad k \in \mathbb{Z}^n \setminus \{0\} \tag{18}$$

where $0 < c(\varepsilon) \longrightarrow 0$ for $\varepsilon \longrightarrow 0$.

A more quantitative formulation of this theorem and a complete proof of it will appear as part of our work on small divisors in which the Newton method is avoided. Indeed, the classical technique of iteration of celestial mechanics is used. But instead of linearization at each step of the iteration process we construct a fixed point of a certain functional operator which simultaneously gives a generating function as well as the canonical transformation belonging to it. Theadvantage of this method for small divisor problems can be compared with the Picard-Lindelöf-iteration as the dominating method for proving the existence of solutions of ordinary differential equations whereas the Newton method is used at most fo numerical purposes.
It has often been said that the rapid convergence of the Newton iteration is necessary for compensating the influence of small divisors like those in (18). But a deeper analysis shows that this is not true. Also the results of Aubry [**1**] ,
Hermann [**5**], and Mather [**7**] (see also Katok [**6**]) on special classes of small divisor problems point in this direction. Historically the Newton method was surely necessary to establish the main theorems of the KAM-theory. But for clarifying the structure of small divisor problems the Newton method is not useful because it compensates not only the influence of small divisors but also many bad estimates veiling the true structure of the problems.
Our method leads, in all analytic soluble small divisor problems, to the natural bound

$$\int_1^\infty \log \frac{1}{M_s} \frac{ds}{s^2} < \infty \tag{19}$$

where

$$M_s = \min_{0 \leq |k| \leq s} |\langle k, \omega \rangle| > 0 \tag{20}$$

and ω is the frequency vector of the problem. Also in (18) the function $s \longrightarrow \exp(-s^{1/2})$ can be replaced by a monotonically decreasing function $s \to M_s$ satisfying (19).
For problems near a singular point condition (19) is equivalent to Bryuno's condition [**2**]. So we cannot obtain more than Bryuno got by means of the Newton method besides of more clearness in the formulae. However, we are able to make the length of the iteration steps arbitrarily small such that in the limit we integrate in the presence of small divisors in the sense that the iteration process can be considered as a Riemannian sum which tends to a Riemann integral, provided (19) is satisfied.
In problems on the torus the situation is more difficult. But using the theory of elliptic functions we are able to show that in the presence of small divisors of constant type, that is we have

$$sM_s \geq \text{const.} > 0 \quad \text{for} \quad s > 0$$

in (20), the length of the iteration steps may also tend to zero, at least in some important examples, and this relates our work to the work [5] of M.R.Herman.

4. Lower Dimensional Invariant Tori

We still consider the more general Hamiltonian system

$$\frac{dx}{dt} = \frac{\partial H}{\partial y}, \quad \frac{du}{dt} = \frac{\partial H}{\partial v}, \qquad \frac{dy}{dt} = -\frac{\partial H}{\partial x}, \quad \frac{dv}{dt} = -\frac{\partial H}{\partial u} \tag{21}$$

where

$$H: T^n \times B \times W \longrightarrow \mathbb{R}$$

is real analytic, $B \subseteq \mathbb{R}^n$ is open and connected, and

$$W = \{ w=(u,v) \in \mathbb{R}^m \times \mathbb{R}^m \mid |w| < r \}$$

is some neighbourhood of $w=0$ in $\mathbb{R}^m \times \mathbb{R}^m$. Moreover H has the form

$$H = H_0(y) + \frac{1}{2} \langle w, wP(y) \rangle + H_2(x,y,w) \tag{22}$$

where $\langle .,. \rangle$ is the usual inner product of $\mathbb{R}^m \times \mathbb{R}^m$ and $P(y)$ is a symmetric 2mx2m-matrix analytically depending on y in B. We assume

$$|H_2(x,y,w)| \leq \varepsilon \quad \text{for all } x \in T^n,\ y \in B,\ w \in W. \tag{23}$$

The unperturbed system ($\varepsilon = 0$) has the invariant tori

$$y = b \in B, \ w = 0$$

with the quasiperiodic trajectories (4). Also in this extended case the question of survival to invariant tori of (21),(22),(23) for small $\varepsilon > 0$ arrise.
An unavoidable condition for the survival seems to be that the purely imaginary eigen values

$$\pm i\Omega_j, \quad \Omega_j : B \longrightarrow \mathbb{R}, \quad j=1,\dots,m_0 \leq m$$

of the w-part of the unperturbed system ($\varepsilon = 0$)

$$\frac{dx}{dt} = \frac{\partial H_0}{\partial y}(y), \quad \frac{dy}{dt} = 0, \quad \frac{dw}{dt} = wP(y)J, \quad J = \begin{pmatrix} 0 & -E \\ E & 0 \end{pmatrix} \tag{24}$$

are different from zero and different from one another:

$$\left.\begin{array}{l}(\Omega_j - \Omega_l)(y) \neq 0 \quad , \quad y \in B \quad , \quad j \neq l \quad , \\ \Omega_j(y) \neq 0 \; ; \quad j,l = 1,\ldots,m_0 \; .\end{array}\right\} \qquad (25)$$

There have appeared papers of Eliasson [3] and Pöschel [9] in which the existence of invariant tori for the perturbed system (21),(22),(23) for small $\varepsilon>0$ are constructed in the case $m_0=m$, that is, only purely imaginary eigen values in (24) do exist. Moreover it is assumed that (10) is valid and then some other conditions are required which we do not specify here because we like to remark that all such additional conditions are superfluous even in the general case $1 \leq m_0 < m$ and that H_0 has only to be nondegenerate.
The reason for this is that besides of (18) the diophantine inequalities

$$|<k,\bar{\omega}(b)> + \bar{\Omega}(b)| \geq c(\varepsilon)e^{-|k|^{\frac{1}{2}}} \qquad \text{for all } k \quad \mathbb{Z}^n \qquad (26)$$

have to be satisfied for sufficiently many $b \in B$ and for $(\bar{\omega},\bar{\Omega})$ sufficiently near by (ω,Ω) where $\Omega:B \longrightarrow \mathbb{R}$ is a real analytic function standing for Ω_j or $\Omega_j - \Omega_l$ in (25). Now (26) can be satisfied in the same way as (18) provided that ω/Ω is nondegenerate, and this is the case if and only if ω is nondegenerate because of (25). So the nondegeneracy of H_0 is the only condition guaranteeing the exstence of invariant tori of the perturbed system (21),(22),(23) for small $\varepsilon>0$.
The main problem in the proof of the theorem above and its extension to lower dimensional invariant tori is the "approximation of dependent quantities" as Sprindzuk [11] calls problems of diophantine approximation with too fiew unknowns. In the theory of dynamical systems there are two problems of this sort : The degeneracy problem for the angular variables, that is, there are not enough free parameters for controlling the frequencies, and the existence of lower dimensional tori, and both problems are transcendentally connected in the analytic case, as we have indicated above.
The first problem is much more difficult than the second one because the implicite function theorem has to be avoided which is the basic tool in KAM-theory to control the frequencies.
In our talks in Moser's seminar in Zürich (January 1986) and at the Number theory and dynamical systems conference in York (April 1987) we mainly treated the second problem because we could still not come through the first one. In the course of the year 1987 we learned to prove the theorem in section 3 for a class of n-tupletts of independent Taylor coefficients (12) until we finally succeeded in our talk at the IVth German-French Meeting on Mathematical Physics in Marseille (March 1988) to prove the existence theorem under the condition of mere nondegeneracy of the unperturbed Hamiltonian.
The extension of this theorem to the problem of lower dimensional tori is straightforward provided only purely imaginary eigenvalues are

present. In the general case new difficulties arrise because the semigroup-approach frequently used in hyperbolic problems is not applicable in the presence of complex frequencies. The same difficulties arrise if one likes to construct the stable and unstable manifold in dissipative systems not by means of real analysis according to Hadamard and Perron but by complex analysis.
In order to handle such problems we studied intensively linear equations containing small divisors produced by complex frequencies. It turned out that better results can be obtained by means of the best approximation of differentiable almost periodic functions in two variables by trigonometrical polynomials than by means of the geometry of numbers counting lattice points. We developped such a theory of best approximation in two variables, not available in the literatur, and pressed it in an identity for the Laplacian in the plane which is verified by evaluating iterated integrals. See our contribution to the volume on the occasion of Moser's 60th birthday.

5. The Twist Mapping Theorem

Our nondegeneracy conditions for Hamiltonian systems are completely analogous to those sufficient for the existence of invariant curves of area preserving mappings of an annulus. We consider a perturbed twist mapping

$$M : \begin{cases} x_1 = x + h(y) + f(x,y) \\ y_1 = y + \qquad + g(x,y) \end{cases}$$

where

$$f,g : T^1 x B \longrightarrow \mathbb{R} , \quad B =]\alpha , \beta[\quad , \ 0<\alpha<\beta$$

are real analytic functions satisfying the estimate

$$|f(x,y)| + |g(x,y)| \leq \varepsilon \ , \quad \text{for all } (x,y) \in T^1 x B. \tag{27}$$

The function $h: B \longrightarrow \mathbb{R}$ is assumed to be real analytic and to satisfy the condition

$$\frac{dh}{dy}(y) \neq 0 \quad , \quad \text{for all } y \in B. \tag{28}$$

The unperturbed mapping ($\varepsilon = 0$) has invariant circles $y_1 = y = b \in B$.

Moser [10] has proved that many of these invariant circles survive as invariant curves of the mapping M with $\varepsilon > 0$ sufficiently small in (27) provided M is not necessarily area preserving but has the intersection property. This means that simply closed curves in $T^1 x B$ near circles

$T^1 x \{b\}$, $b \in B$ intersect its image under M.

The condition (28) corresponds to condition (10) for Hamiltonian systems. Nondegeneracy of h in the broadest sense corresponding to our considerations above means here: $\frac{dh}{dy}$ must not vanish identically in B, that is locally, at each point $b \in B$ the Taylor expansion

$$h(y) = \sum_{j=0}^{\infty} \frac{h^{(j)}(b)}{j!}(y-b)^j \quad , \quad h^{(j)}(b) \neq 0 \ , \quad j=j(b) \geq 1$$

has at least one non vanishing coefficient $h^{(j)}(b) \neq 0$ for some $j=j(b) \geq 1$. This situation can easily be reduced to the case (28) as Moser has shown: One introduces a new radial variable

$$z = h(y) \ ,$$

and this equation can be locally inverted. The intersection property remains valid under such transformations.
In general the twist mapping theorem has been formulated for the existence of individual invariant curves. But it is not difficult, to give also a measure theoretic version similar to the theorem in section 3 using the same tools as for the individual case.
We conclude with the remark that similar to the considerations of this paper the condition of nondegeneracy can be weakened in the case of invariant tori lying on an energy surface and in the case of Arnol'd's theorem [4,p185] for applications in celestial mechanics.

6. References

[1] Aubry, S. and Le Daeron, P.Y., The discrete Frenkel-Kontorova model and its its extensions I, Physica 8D, 381-422, 1983

[2] Bryuno, A.D., Analytic Form of Differential Equations, Trudy MMO 25(1971), 119-262 = Translations of the Moscow Math. Soc. 25(1971), 131-288

[3] Eliasson, L.H., Perturbations of Stable Invariant Tori, Preprint 1985

[4] Encyclopaedia of Mathematical Sciences, Vol. III, Springer-Verlag, 1988

[5] Hermann, M.R., Sur les Courbes Invariantes par les Difféomorphismes de l'anneau, Volume I, Astérisque 103-104

[6] Katok, A., Some remarks on Birkhoff and Mather twist map theorems, Ergodic Theory and Dynamical Systems 2, 185-194, 1982

[7] Mather, J.N., Existence of quasi-periodic orbits for twist homeomorphismus of the annulus, Topology 21, 457-467, 1982

[8] Moser, J., Stable and Random Motions in Dynamical Systems, Annals of Mathematics Studies Nr. 77, Princeton University Press 1973

[9] Pöschel, J., On Elliptic Lower Dimensional Tori in Hamiltonian Systems, Preprint 1988, Universität Bonn

[10] Siegel, C.L.-Moser, J.K., Lectures on Celestial Mechanics, Springer-Verlag 1971

[11] Sprindzuk, G.V., Metric Theory of Diophantine Approximations, John Wiley & Sons 1979

ENERGY FORMS IN TERMS OF WHITE NOISE

Ludwig Streit
UM - Area de Matematica
Braga, Portugal
and
BiBoS - Univ. Bielefeld
Bielefeld, FRG

Abstract: Local Dirichlet forms have developped into an important tool for the non-perturbative treatment of quantum dynamics and diffusion processes. With a view towards quantum field theory it is desirable to extend the formalism to infinite dimensional ("configuration") spaces. As a starting point we choose the L^2 space (L^2) over the White Noise measure. More general measures are constructed by first embedding (L^2) into a triple $(\mathcal{S}) \subset (L^2) \subset (\mathcal{S})^*$. Positive elements of $(\mathcal{S})^*$ are measures. We call a positive generalized functional admissible if the corresponding measure ν gives rise to a closable energy form on $(L^2)_\nu$. Any admissible generalized functional gives rise to a Markovian form in the sense of Fukushima.

I.- Introduction. - My first contact with Dirichlet forms and their use in quantum dynamics dates back to 1976 when Raphael Hoegh-Krohn visited Bielefeld and I was privileged to join a collaboration which has not only brought a rich harvest of scientific results[1] but as well the experience of good friendship in the years since.

As a result of this joint work, "energy forms"

$$\epsilon\,(f) = \int (\nabla f(x))^2\ \psi^2(x)\ d^n x$$

turned out to be an extremely effective tool for the description of very singular interactions in quantum mechanics, since

$$\epsilon\,(f) = (f,\, H\, f)_{L^2(\psi^2 dx)}$$

where the operator $H \simeq -\,\Delta + V(x)$ is a local Schroedinger Hamiltonian with $V = \nabla\psi/\psi$ if

[1]see e.g. refs. 3-5

S. Albeverio et al. (eds.), Stochastics, Algebra and Analysis in Classical and Quantum Dynamics, 225–233.

this perturbative expression makes sense. The point of introducing ϵ is that the energy form is valid as a non-perturbative definition of the Hamiltonian H much more generally, i.e. for interactions much too singular for a perturbative treatment.

At the same time the properties of the energy form guarantee that H is the generator of a Markov semigroup[2]. We can associate with ϵ a diffusion process, the "Euclidean" quantum theory.

In this framework the perturbative $H = H_0 + V$ definition of dynamics is avoided, instead the dynamical information resides in the function ψ which physically is nothing but the "ground state wave function". In this sense for quantum field theory this approach offers a correct version of the old proposal by Coester and Haag[3] to formulate quantum dynamics in terms of the vacuum, se also[4].

Of course the main difficulty lies in the transition from an n-dimensional configuration space $\mathbb{R}^n$ to infinitely many degrees of freedom. No "flat" integration measure such as $\int d^\infty x$ is available, - but White Noise measure μ suggests itself as a candidate, and we might try

$$\epsilon(F) = \int \int (\partial_s F[x])^2 ds\ \Phi[x]\ d\mu[x].$$

Indeed the simplest case is by now well studied:

$$\Phi = 1$$

with

$$H = \int ds\ \partial_s{}^*\ \partial_s = N = -\Delta_\infty$$

synonymously the number operator, "Laplace-Beltrami operator", and generator of an infinite dimensional Ornstein-Uhlenbeck process, an ultra-local field theory.
But more general Φ should be studied and have been. Takeda[5] for example considers certain positive

$$\Phi \in L^1(d\mu)$$

so that $d\nu = \Phi d\mu$ is absolutely continuous with respect to the White Noise measure μ. However experience from quantum field theory underlines the importance of

---

[2]ref. 8

[3]ref. 7

[4]refs. 2,6.

[5]ref. 15

$$\epsilon(F) = \int d\nu \, (\nabla F)^2$$

with ν singular to μ. It is this situation that we have been addressing[6]. Considering generalized Radon-Nikodym derivatives is relatively easy in the finite dimensional case where positive generalized functions are known[7] to define measures. Here, in the infinite dimensional case, we must first prepare this kind of result.

II. - **Generalized White Noise Functionals** - The first step must be to decide on suitable spaces of test and of generalized functionals, i.e. on a Gelfand triple

$$E \subset (L^2) \subset E^*.$$

Of course this is if anything even less unique than in the finite dimensional case, and specific choices will be dictated by specific needs.

Here we proceed as follows. Consider the Fock (or multiple Wiener integral) decomposition of

$$(L^2) \simeq \bigoplus_n \mathrm{Sy}\; L^2(\mathbb{R}^n, n!\, d^n s).$$

Thus with any operator A on $L^2(ds)$ we can associate its "2nd quantization"[8]

$$\Gamma(A) \simeq \bigoplus_n A^{\otimes n}.$$

We shall consider self-adjoint operators A and have

$$\Gamma(A^p) = \Gamma(A)^p$$

and for example if

$$F[x] = e^{i(x,f)}$$

we find

$$\Gamma(A^p)F[x] = \text{const. } e^{\,i(x,A^p f)}.$$

[6]ref. 10

[7]ref.9, p.128 ff.

[8]ref. 14, p.309

Now we choose in particular

$$A = 1+s^2 - \frac{d^2}{ds^2}$$

i.e. the harmonic oscillator Hamiltonian:

$$A\, e_k = (2k+2)\, e_k,$$

where $e_k(s)$ are Hermite functions. We denote by $(\mathcal{S}_p)$ the domain of $\Gamma(A^p)$,

$$D\,(\Gamma(A^p)) = (\mathcal{S}_p) \subset (\,L^2)$$

and have

$$\ldots (\mathcal{S}_p) \subset (\mathcal{S}_{p-1}) \subset \ldots \subset (\,L^2) \subset \ldots \subset (\mathcal{S}_{-p}) \subset \ldots\,.$$

Defining

$$(\mathcal{S}) = \bigcap_p (\mathcal{S}_p)$$

and endowing $(\mathcal{S})$ with the projective limit topology, we consider the triple

$$(\mathcal{S}) \subset (\,L^2) \subset (\mathcal{S})^*.$$

Note that

$$(\mathcal{S})^* = \bigcup_p (\mathcal{S}_{-p}).$$

Furthermore we have the following results

<u>LEMMA 1:</u> $(\mathcal{S})$ is dense in (L^2).
Proof: For any $f \in \mathcal{S}$, $F[x] = e^{i(x,f)} \in (\mathcal{S})$, and the span of such F is dense in (L^2).

<u>LEMMA 2:</u> $(\mathcal{S})$ is a $*$-algebra.
The proof[9] amounts to verification of an estimate

$$\|F\,G\|_{2,p} \leq \|F\|_{2,p+q}\,\|G\|_{2,p+q}\,,$$

similar estimates for monomials have already been formulated[10]. Essential for the argument is the product formula for normal ordered monomials and the estimate

$$\Gamma(A^p) \geq 2^{pN}.$$

LEMMA 3: (Kubo, Yokoi[11]): Any test functional $F\in(\mathcal{S})$ has a version

$$\tilde{F}[x] = \sum_n \left\langle :x^{\otimes n}:, F_n \right\rangle \quad \text{with} \quad F_n \in \mathrm{Sy}\,\mathcal{S}(\mathbb{R}^n),$$

well-defined for any $x\in\mathcal{S}^*(\mathbb{R})$.

Conversely, $F\in(\mathcal{S})$ iff

$$\sum_n n!\,\left\| (A^p)^{\otimes n}\,F_n \right\|_2^2 < \infty \quad \text{for all } p\geq 0.$$

We shall use this version to extend functionals beyond the support of μ, omitting the indication $\tilde{}$ to simplify notation. On the Fock representation of $(\mathcal{S})$ the annihilation operators a_f are well defined for any $f\in\mathcal{S}(\mathbb{R})$. We denote by ∂_f their image in (L^2).

LEMMA 4: $\partial_f : (\mathcal{S}) \to (\mathcal{S})$.

Proof: Using Lemma 3 it suffices to verify an estimate of the form

$$\left\| \partial_f\,F \right\|_{2,p} \leq \sqrt{n}\,2^{-n}\,\left\| A^{-1}f \right\|_2 \left\| F_n \right\|_{2,p+1}.$$

For more details on the White Noise calculus we refer to the contribution of J.Potthoff in this volume.

Here we introduce an infinite dimensional generalization of the gradient operator with a view towards its use in energy forms:

---

[9] ref. 10, Appendix

[10] ref. 12

[11] ref. 13

DEFINITION: $\nabla : (L^2) \to (L^2) \otimes l^2$

$$\nabla : F \mapsto (\partial_{e_k} F)_{k=0}^{\infty}$$

and

$$(\nabla F)^2 = \sum_k (\partial_{e_k} F)^2.$$

With these concepts we can establish

LEMMA 5: $F \in (\mathcal{S}) \Rightarrow (\nabla F)^2 \in (\mathcal{S})$

Proof: Use Lemma 4 regarding the derivative and proceed as in Lemma 2 regarding products and sums.

Finally we introduce positive cones in the triple as follows.

DEFINITION: $(\mathcal{S})_+ = \{ F \geq 0 \} \in (\mathcal{S})$

$$(\mathcal{S})^*_+ = \{ \Phi \in (\mathcal{S})^* : (\Phi, F) \geq 0 \text{ if } F \in (\mathcal{S})_+ \}.$$

For positive generalized functionals $\Phi \in (\mathcal{S})^*_+$ we have, as in the finite dimensional case, a representation in terms of measures. It is given in Yokoi's[12]

THEOREM: For any positive generalized functional $\Phi \in (\mathcal{S})^*_+$ there is a unique finite positive measure ν on the Borel algebra over the space of tempered distributions with

$$(\Phi, F) = \int F \, d\nu.$$

In other words the positive cone in $(\mathcal{S})^*$ offers a natural unifying framework for the construction of measures which live on the same sigma-algebra but may fail to be absolutely continuous with respect to White Noise. The elements of $(\mathcal{S})^*_+$ play the role of generalized Radon-Nikodym derivatives. Now we are tooled up to discuss infinitely dimensional Dirichlet forms.

---

[12] ref. 16

III.- Energy Forms in Terms of White Noise. - Guided by the experience from the finite dimensional case we shall not expect that all positive distributions give rise to (closable) energy forms. Hence we make the following

DEFINITION: Consider $\Phi \in (\mathcal{S})^*_+$ and the measure ν corresponding to it. If

$$\epsilon(F) = \int d\nu \, (\nabla F)^2 \quad \text{with } D(\epsilon) = (\mathcal{S})$$

is closable on $L^2(d\nu)$ we shall call ϵ an "energy form" and Φ "admissible"[13].

As is well-known we obtain a self-adjoint, positive "Hamiltonian" operator canonically whenever we construct an energy form:

THEOREM (Kato[14]): To every energy form ϵ there corresponds a unique self-adjoint positive operator H in $L^2(d\nu)$ such that

$$\epsilon(F) = \| H^{1/2} F \|^2 \text{ and } D(H^{1/2}) = D(\overline{\epsilon})$$

Furthermore, again as in the finite dimensional case, we can show[15] that any such form is Markovian:

THEOREM: For any admissible distribution Φ the corresponding energy form is Markovian in the sense of Fukushima[16].

The simplest case of course is obtained by considering $\Phi=1$ as mentioned in the Introduction. But of course our main interest lies with examples of admissible distributions that lead out of the equivalence class of the White Noise measure. The following construction produces positive functionals that are, in general, distributions:

Example: Let K be a self-adjoint operator in $L^2(\mathbb{R})$, bounded below by $-1+\epsilon$ for some $\epsilon>0$. Under this condition we may define the "normalized exponential"

$$\text{Nexp}\left(-\tfrac{1}{2}(x.Kx)\right) \in (\mathcal{S})^*_+$$

---

[13]Implicit in this requirement is the condition that ϵ respects ν-classes.

[14]ref. 11, p.331, cf. also ref. 4

[15]ref. 10

[16]ref. 8

by its S-transform $\exp\left(-\frac{1}{2}(f.\widetilde{K}f)\right)$, where $\widetilde{K}= K\cdot(K+1)^{-1}$. The renormalization involved in this definition will in general be infinite, but the semiboundedness condition suffices to ensure that the functional is in $(\mathcal{S})_+^*$.

To decide the admissibility of such generalized functionals we cite the following

THEOREM: Φ is admissible if for all k

$$\partial_{e_k}\Phi = B_k\Phi \quad \text{for some} \quad B_k \epsilon (\mathcal{S}).$$

Here derivation and multiplication are defined by duality since they are well-defined for test functionals. The theorem is proved by noting that forms such as $\epsilon(F)=||\nabla F||^2$ are closable if ∇ is. To show this latter fact it sufices to demonstrate that the adjoint of ∇ is densely defined, which in turn follows easily from the condition posed in the theorem.

As an immediate application we note that for the functional

$$\Phi[x] = N\exp\left(-\tfrac{1}{2}(x.Kx)\right) \in (\mathcal{S})^*_+$$

$$B_k = -(x,Ke_k),$$

so that it suffices to postulate $Ke_k \in S(\mathbb{R})$ $\forall k \epsilon \mathbb{N}$ to ensure admissibility.

IV.- Conclusions. - White Noise Analysis offers a paricularly useful framework for infinite dimensional analysis and especially for the generalization of energy form theory to infinite dimensions as suggested by quantum field theory. The construction of smooth and generalized random variables via the second quantization of the harmonic oscillator Hamiltonian produces a Gelfand triple with many desirable properties, we emphasize here in particular Yokoi's Theorem. With its help the positive cone of the distribution space offers a rich source of measures; the admissible elements then give rise to dynamics of quantum (resp. Markov) fields, a setup which warrants further exploration[17].

[17]Since these lectures were delivered the White Noise Analysis framework as presented here has passed an important test; in ref.1 we show that it encompasses the highly non-trivial models of constructive quantum (and Markov) field theory.

References:

1.- S. Albeverio, T. Hida, J. Potthoff, L. Streit: "The Vacuum of the Hoegh-Krohn Model as a Generalized White Noise Functional". Phys. Lett. B. 217, 511 (1989).

2.- S. Albeverio, R. Hoegh-Krohn: "Diffusion Fields, Quantum Fields, andFields with Values in Lie Groups ". In "Stochastic Analysis and Applications", M. Pinsky, ed. - M. Dekker, New York, 1984.

3.- S. Albeverio, M. Fukushima, W. Karwowski, R. Hoegh-Krohn, L. Streit: "Capacity and Quantum Mechanical Tunneling". Comm. Math. Phys. 81, 501 (1981).

4.- S. Albeverio, R. Hoegh-Krohn, L. Streit: "Energy Forms, Hamiltonians, and Distorted Brownian Paths ". J. Math. Phys. 18, 907 (1977).

5.- S. Albeverio, R. Hoegh-Krohn, L. Streit: "Regularization of Hamiltonians and Processes". J. Math. Phys. 21, 1636 (1980).

6.- H. Araki: "Hamiltonian Formalism and the Canonical Commutation Relations in Quantum Field Theory". J. Math. Phys. 1, 492 (1960).

7.- F. Coester, R. Haag: "Representation of States in a Field Theory with Canonical Variables". Phys. Rev. 117, 1137 (1960).

8.- M. Fukushima: "Dirichlet Forms and Markov Processes". North Holland - Kodansha 1980.

9.- I. M. Gelfand, N. J. Vilenkin: "Verallgemeinerte Funktionen", Bd. IV. VEB Verlag der Wissenschaften, Berlin, 1964.

10.- T. Hida, J. Potthoff, L. Streit: "Dirichlet Forms and White Noise Analysis". Comm. Math. Phys. 116, 235 (1988).

11.- T. Kato: "Perturbation Theory for Linear Operators". Springer, Berlin, 1966.

12.- I. Kubo, S. Takenaka: "Calculus on Gaussian White Noise II". Proc. Japan Acad. 56 Ser. A, 411 (1980).

13. I. Kubo, Y. Yokoi: "A Remark on the Space of Testing Random Variables in the White Noise Calculus". Preprint 1987.

14.- M. Reed, B. Simon: "Functional Analysis", vol. I. Academic Press, New York, 1972.

15.- M. Takeda: "On the Uniqueness of the Markovian Self-Adjoint Extensions of Diffusion Operators on Infinite Dimensional Spaces". Osaka J. Math. 22, 733 (1985).

16.- Y. Yokoi: "Positive Generalized Brownian Functionals". Kumamoto preprint, 1987.

LINEAR STABILITY AND THE PARAMETER MODULATION APPROACH FOR SOLITARY WAVES IN HAMILTONIAN SYSTEMS WITH SYMMETRIES

Joachim Stubbe
Fakultät für Physik, Theoretische Physik
Universität Bielefeld
D-4800 Bielefeld 1, RFA

1. Introduction

I have studied abstract Hamiltonian systems of the form

$$\frac{du}{dt} = J\, E'(u) \tag{H}$$

which are locallly well-posed in a real Hilbert space X, where E denotes the energy functional on X, E′ its Frechet-derivative and J is a skew-symmetric linear operator. I assume that the system (H) is invariant under a representation T(·) of the group G = (ℝ,+). By a solitary wave I mean a solution of (H) of the form

$$u(t) = T(\omega t)\,\phi_\omega\,, \quad \phi_\omega \in X \tag{1.1}$$

and I assume that such solutions exist for an interval of parameters ω.

Typical examples of systems exhibiting solitary wave solutions (or solitons) are nonlinear wave equations

$$\psi_{tt} - \psi_{xx} = h(\psi) \tag{NLW}$$

for (real) fields ψ with solitary waves of the form $\psi_\omega(x-\omega t)$, or nonlinear Schrödinger equations

$$iu_t + u_{xx} + g(|u|^2)u = 0 \tag{NLS}$$

S. Albeverio et al. (eds.), Stochastics, Algebra and Analysis in Classical and Quantum Dynamics, 235–246.

with solitary waves $\phi_\omega(x)\exp(i\omega t)$, etc.

In the applications a basic requirement on solitary waves is the condition of stability. For the stability analysis many different notions of stability have been used in the literature, e.g.

Nonlinear (orbital) stability: In view of the invariance one considers the stability of the ϕ_ω-orbit

$$O_\omega \equiv \{T(s)\,\phi_\omega,\ s \in \mathbb{R}\}. \tag{1.2}$$

The ϕ_ω-orbit is called stable if for any given tubular neighborhood U of O_ω there exists a tubular neighborhood V of O_ω such that all solutions u(t) of (H) with initial data u(O) in V exist for all t $\geq$ O and remain in U.

Energetic stability: Associated to the symmetry there is another conserved functional Q. The (time-dependent) vector ϕ_ω is a critical point of

$$L_\omega \equiv E - \omega\,Q. \tag{1.3}$$

Now one says that ϕ_ω is energetically stable, if it (locally) minimizes the energy E subject to constant Q.

Linear stability: If I linearize the system (H) around a solitary wave of the form (1.1), I shall obtain the linear Hamiltonian system

$$\frac{dw}{dt} = J\,H_\omega\,w \tag{H_{lin},ω}$$

where H_ω is the "linearized Hamiltonian" given by

$$H_\omega \equiv L''_\omega(\phi_\omega). \tag{1.4}$$

Now, roughly spoken, ϕ_ω is called linearly stable if any solution w of (H_{lin},ω) remains bounded. (This is equivalent to the Liapunov-stability of the trivial solution.)

In a recent paper, Grillakis, Shatah and Strauss find sharp conditions for the orbital stability of solitary waves under rather general assumptions [1]. The basic assumption is a condition on the critical point ϕ_ω of L_ω which can be expressed as a property of the linearized operator: Let H_ω have at most one negative eigenvalue. Now, the invariance implies that zero is also in the spectrum of H_ω. The rest of the spectrum has to be positive and bounded away from zero.

In the following I shall only discuss the more interesting case where H_ω is one negative eigenvalue since it will show how a symmetry can "kill" an unstable direction. Using this assumption, Grillakis, Shatah and Strauss show that the ϕ_ω-orbit is stable if and only if the scalar function

$$d(\omega) \equiv L_\omega(\phi_\omega) \tag{1.5}$$

is convex at ω. The main intermediate step is to show that ϕ_ω is energetically stable if and only if $d(\omega)$ is convex at ω. Hence, under the above assumption on the linearized operator the nonlinear stability and the energetic stability are equivalent.

The first aim of this contribution is to report on my recent work on the linear stability problem which I could solve under the same assumption on H_ω provided $d''(\omega)$ is nonzero [2]. The main difficulty is the following: Obviously the definition of linear stability cannot hold for (H_{lin},ω) since the critical point around which we linearize is degenerate. To explain this fact we observe that the functions

$$\psi(t,\theta,\omega) \equiv T(\omega t + \theta)\,\phi_\omega \tag{1.6}$$

form a two-parameter family of solitary wave solutions of (H). Formally, the derivatives of ψ with respect to the parameters θ and ω are solutions of the linearized system which (may) grow linearly in time. These "secular modes" exist for all solitary waves (independently of all other stability properties). In [2] I constructed a subspace W of X (with codimension two) where the secular modes are removed. This space W is invariant for the linear evolution equation and my main result is then, that the linear stability of ϕ_ω under perturbations in W is equivalent to its orbital (resp. energetic) stability. Finally, I want to propose how my linear stability analysis can be used to prove results on the structural stability of solitary waves of the form (1.1). In the physics of solitary waves there is the important question how the solitary waves change in the presence of external fields. Now the perturbation appears in the evolution equation (H) and not alone in the initial data. To be more precise, I shall consider the perturbed system

$$\frac{du}{dt} = J\,E'(u) + \varepsilon\, f(u) \qquad (H_\varepsilon)$$

where ε is a small parameter and f is a smooth mapping in X. One way physicists treat this equation is the so-called parameter modulation approach which has been performed for concret problems as nonlinear wave equations [3] or nonlinear Schrödinger equations [4]. The idea is that the main effect of the perturbation is a change ("modulation") of the solitary wave parameters. (See [3,4] or Section 3 for the details). Here I want to give a mathematical justification of this approach which even holds in the abstract framework mentioned above.

2. Linear stability in the presence of symmetry

First of all I have to give some technical preliminaries (see [2] for more details).

Let X be a real Hilbert space with scalar product (,). By X^* I denote its dual and by $<\ ,\ >$ the dual pairing between X and X^*. There is an isomorphism $I : X \to X^*$ such that $<Iu,v> = (u,v)$ for all u, v $\in$ X.

I assume that J is a skew-symmetric linear operator from X^* onto X, i.e.

$$<u^*, Jv^*> = -<Ju^*, v^*> \qquad u^*, v^* \in X^*. \qquad (2.1)$$

The energy $E : X \to \mathbb{R}$ is assumed to be of class C^2 where the first derivative is written as $<E'(u),v>$ and its second derivative as $<E''(u)w,v>$. Let $T(\cdot)$ be a one parameter group of operators on X with infinitesimal generator $T'(0)$.

I assume that the energy E and the system (H) is invariant under $T(\cdot)$.

The first condition is $E(u) = E(T(s)u)$, the latter can be guaranted by the condition

$$T(s)\,J = J\,T^{*}(-s) \tag{2.2}$$

where $T^{*}(s) : X^{*} \to X^{*}$ is the adjoint of $T(s)$.

In order to construct the conserved quantity Q associated to $T(\cdot)$ I assume that there exists a bounded linear operator $B : X \to X^{*}$ such that

$$JB \text{ is an extension of } T'(0) \tag{2.3a}$$

$$B = B^{*}, \text{ i.e. } \langle Bu,v\rangle = \langle Bv,u\rangle\ . \tag{2.3b}$$

Then Q can be defined by

$$Q(u) = \tfrac{1}{2}\langle Bu,u\rangle. \tag{2.4}$$

Now I consider solitary waves of the form (1.1), i.e.

$$u(t) = T(\omega t)\,\phi_{\omega}$$

for an interval of parameters ω. It is reasonable to assume that $T'(0)\,\phi_{\omega} \neq O$. ϕ_{ω} is then a critical point of $L_{\omega} = E - \omega Q$, i.e.

$$L'_{\omega}\,(\phi_{\omega}) = O\ . \tag{2.5}$$

If I write a solution u(t) of (H) as $u(t) = T(\omega t)\,[\phi_{\omega} + w(t)]$ and make the first order approximation I obtain the linearized Hamiltonian system

$$\frac{dw}{dt} = J\,H_{\omega}\,w \tag{H_{lim},ω}$$

where $H_{\omega} = L''_{\omega}(\phi_{\omega})$. I use the "spectral assumption" of [1]:

Let H_{ω} have exactly one negative (simple) eigenvalue.

The nullspace of H_ω, $N(H_\omega)$, in only spanned by $T'(0)\phi_\omega$. (The equation $H_\omega T'(0)\phi_\omega = 0$ follows immediately from the assumptions made above.) The rest of the spectrum is positive and bounded away from zero.

The energy functional of the linearized system is given by the quadratic form

$$K_\omega(w) = \frac{1}{2} < H_\omega\ w, w > . \tag{2.6}$$

The secular modes associated to the solitary wave family (1.6) are represented by the following explicit solutions of the linearized system.

$$w_1(t) = T'(0)\ \phi_\omega \tag{2.7a}$$

which follows from the assumption on $N(H_\omega)$ and

$$W_2(t) = t \cdot T'(0)\ \phi_\omega + \phi'_\omega \ , \quad \text{with} \quad \phi'_\omega = \frac{d\phi_\omega}{d\omega} \tag{2.7b}$$

which follows from differentiating (2.5) with respect to ω (i.e. $H\phi'_\omega = B\phi_\omega$).

Next I define the space W on which I shall constrain the linear evolution. For a linear operator A set

$$Ng(A) = \bigcup_{n=1}^{\infty} N(A^n)$$

the 'generalized nullspace' of A. I define W by the formula

$$W = X \cap [Ng((JH_\omega)^*)]^{\perp} \tag{2.8}$$

where '$\perp$' denotes orthogonality with respect to the dual pairing $< , >$. I define linear stability of ϕ_ω as follows:

Definition: ϕ_ω is linearly stable if for all $\varepsilon > 0$ there exists $\delta > 0$ having the following property: If $w_0 \in W$ and $\|w_0\| < \delta$ and $w(t)$ is a solution of the linearized equation with $w(0) = w_0$. Then

$$\sup_{t>0} \|w(t)\| < \varepsilon.$$

Otherwise ϕ_ω is linearly unstable.

My result is (see also [2]) the following:

Theorem: Let $d(\omega) \equiv L_\omega(\phi_\omega)$ with $d''(\omega) \neq 0$. Then the following two conditions are equivalent:

(a) $d''(\omega) > 0$

(b) ϕ_ω is linearly stable.

The proof consists of two main intermediate steps. The first is the construction of the space W.

Lemma 2.1. Let $d''(\omega) \neq 0$. Then

(a) $Ng(JH_\omega) = N(JH_\omega) \cup N((JH_\omega)^2)$

(b) $Ng((JH_\omega)^*) = N((JH_\omega)^*) \cup N((JH_\omega)^{*^2})$.

$Ng(JH_\omega)$ and $Ng((JH_\omega)^*)$ are spanned by the following biorthogonal sets

$$e_1 = T'(0)\,\phi_\omega\,, \quad e_2 = \phi'_\omega \tag{2.9a}$$

$$f_1 = J^{-1}\,\phi'_\omega\,, \quad f_2 = -\,B\phi_\omega \tag{2.9b}$$

where

$$<f_i,e_j> = d''(\omega)\,\delta_{i_j}\,. \tag{2.10}$$

To prove lemma 2.1 one simply solves the equations $(JH_\omega)^n\,u = 0$ (resp. $(H_\omega J)^n u^* = 0$). The finiteness of the construction comes from the fact that $d''(\omega) \neq 0$ (see [2] for the details). □

The second step is to use the energetic stability result of [1] in a somewhat stronger form:

Lemma 2.2:

(a) Let $d''(\omega) > 0$ then there exist constants $c_1, c_2 > 0$ such that

$$c_1 \|w\|^2 \leq \langle H_\omega w,w\rangle \leq c_2 \|w\|^2$$

for any $w \in W$.

(b) Let $d''(\omega) < 0$ then there exists $z \in W$ such that

$$\langle H_\omega z,z\rangle < 0.$$

In the case $d''(\omega) > 0$ the linear stability is now a consequence of the conservation of the linearized energy (2.6). If $d''(\omega) < 0$ one easily constructs a Liapunov-functional which increases along the trajectories of solutions with negative linearized energy.

Remark 2.3: By the construction of W we have $X \simeq W \oplus Ng(JH_\omega)$. $Ng(JH_\omega)$ is the space of the secular evolution (of dimension two). Hence W is in fact the space where the secular modes are removed.

3. The parametermodulation approach in abstract Hamiltonian system

In this section I apply the linear stability analysis of Section 2 to study the perturbed system

$$\frac{du}{dt} = J E'(u) + \varepsilon f(u), \quad 0 < \varepsilon \ll 1 \qquad (H_\varepsilon)$$

locally around a solitary wave of the unperturbed system. Here $f : X \to X$ is a smooth mapping and for simplicity I assume that f behaves under the action of $T(\cdot)$ as follows: $f(T(s)u) = T(s) f(u)$. In this abstract framework the parameter modulation approch may be described as follows. The unperturbed system admits a two parameter family of solitary wave solutions given by (1.6) or equivalently by

$$\psi(t,\Delta(t),\Theta) = T(\Delta(t) + \Theta)\psi_\omega \tag{3.1}$$

where $\frac{d\Delta}{dt} = \omega$. For the perturbed system I expand a solution $u^\varepsilon(t)$ as

$$u^\varepsilon(t) = T(\Delta+\Theta)[\phi_\omega + \varepsilon w(t) + \ldots] \tag{3.2}$$

where $\Delta \equiv \Delta(\varepsilon t)$, $\omega \equiv \omega(\varepsilon t)$ and $\Theta \equiv \Theta(\varepsilon t)$ are not constants but slowly varying functions due to the response of the solitary wave to the external perturbation. For consistency, Δ is defined by the differential equation

$$\frac{d\Delta}{dt} = \omega\,, \quad \Delta(0) = 0.$$

The first order approximation (in ε) leads to

$$\begin{aligned}\frac{dw}{dt} &= J\,H_\omega\, w + f(\phi_\omega) - \dot{\Theta}\, T'(0)\,\phi_\omega - \dot{\omega}\,\phi'_\omega \\ &\equiv J\,H_\omega\, w + \mathfrak{F}_{eff}\end{aligned} \tag{3.3}$$

where $\mathfrak{F}_{eff}$ denotes the "effective source". "·" represents the derivation with respect to $\tau \equiv \varepsilon t$ and $\phi'_\omega = \partial\phi_\omega/\partial\omega$.
By the result of the last section I conclude that if $\mathfrak{F}_{eff} \in W$ and $W(0) = W_0 \in W$ then $w(t) \in W$ for all $t > 0$. Now $\mathfrak{F}_{eff} \in W$, if

$$<f_i, \mathfrak{F}_{eff}> = 0 \qquad i = 1, 2 \tag{3.4}$$

with f_i given in (2.9b). After some simple computations I obtain the so-called modulation equations

$$<J^{-1}\,\phi'_\omega, f(\phi_\omega)> = \dot{\Theta}\, d''(\omega) \tag{3.5a}$$

$$-\ <B\,\phi_\omega, f(\phi_\omega)> = \dot{\omega}\, d''(\omega). \tag{3.5b}$$

If these equations possess a solution (Θ,ω) (uniformly bounded for finite times) then I can constrain the evolution to W. If in addition $d''(\omega) > 0$ for ω obtained by (3.5) then the linear stability result of Section 2 applies. Therefore let $d''(\omega) > 0$. For simplicity I consider only the initial value $W_0=0$ (this choice represents the solitary wave solution of the unperturbed system as initial datum). Writing $\Omega(t) = \exp(tJH_{\omega_0})$ the solution w(t) may be written as

$$w(t) = \Omega(t) \int_0^t \Omega(-s)\, \mathfrak{F}_{eff}(\phi_\omega, \varepsilon s)\, ds \; . \tag{3.6}$$

By construction $w(t) \in W$ and by Lemma 2.2 I can bound w(t) by its linearized energy, i.e. there exists a positive constant c such that

$$\|\varepsilon w\|^2 \leq c \, \langle H_{\omega_0} \varepsilon w, \varepsilon w \rangle \; . \tag{3.7}$$

Now, since $\mathfrak{F}_{eff}$ is almost constant, an extension of the mean ergodic theorem [4] implies that the right hand side of (3.7) tends to zero as ε bends to zero. More precisely, for any $t_0 > 0$ one has

$$\sup_{0<t<t_0/\varepsilon} \|\varepsilon w\| = 0(\varepsilon).$$

This situation can be summarized in the following theorem.

<u>Theorem 2:</u> Expand a solution $u^\varepsilon(t)$ of (H_ε) in the form (3.2). Let the parameter $\Theta = \Theta(\varepsilon t)$, $\omega = \omega(\varepsilon t)$ satisfy the modulation equations (3.5a,b) and let $d''(\omega)$ so for all t. Then, for any $t_0 > 0$, the solution w(t) of the linearized system (3.3) with initial value $w(0) = 0$ statisfies

$$\lim_{\varepsilon \to 0} \sup_{0<t<t_0/\varepsilon} \|\varepsilon W(t)\| = 0.$$

<u>Remark 3.1.:</u> A more detailed result for a larger class of perturbations will follow along the same lines and will be published elsewhere. A basic condition for the application of the ergodic theorem is that the effective force $\mathfrak{F}_{eff}$ is

"almost" constant. The theorem is not applicable for quickly oscillating external forces or to stochastic perturbations. However, I believe that the parameter modulation approach is also a useful tool to treat the problem of stochastic perturbations (Formally, the modulation equations are stochastic differential equations in this case.).

Finally, I want to present two examples of perturbations f(u):

<u>Example 3.1.:</u> Let $f(u) = J\,F'(u)$, i.e. f(u) is a Hamiltonian perturbation having the same properties as $J\,E'(u)$. This example is important since in the applications the nonlinear term $E'(u)$ is only an approximation and $F'(u)$ may be viewed as "higher order" terms. The modulation equations become (let $d''(\omega) > 0$)

$$d''(\omega)\,\dot{\Theta} = -\,\partial F(\phi_\omega)/\partial\omega \tag{3.8a}$$

$$\dot{\omega} = 0. \tag{3.8b}$$

Hence $\omega = \omega_0$ and $\Theta = \Theta_0 + \varepsilon d''(\omega)^{-1}\frac{\partial F(\phi_\omega)}{\partial\omega}\cdot t$ which shows that the solitary wave changes its "velocity" along its orbit.

<u>Example 3.2.:</u> I reconsider the linear stability problem of Section 2, i.e. $f(u)=0$, but perturbations in the initial data: $w(0) = w_0 \neq 0$. Setting $z = w - w_0$ I obtain an equation of the form (3.3)

$$\frac{dz}{dt} = J\,H_\omega z + J\,H_\omega w_0 - \dot{\Theta}\,T'(0)\,\phi_\omega - \dot{\omega}\,\phi'_\omega\,. \tag{3.9}$$

The modulation equations read (again I assume $d''(\omega) > 0$)

$$d''(\omega)\,\dot{\Theta} = -\,\langle B\,\phi_\omega, w_0\rangle \tag{3.10a}$$

$$0 = \dot{\omega}\,. \tag{3.10b}$$

Hence $\omega = \omega_0$ and $\Theta = \Theta_0 - \varepsilon\, d''(\omega)^{-1}\langle B\,\phi_\omega, w_0\rangle\, t$. The meaning of this result is that, to leading order, perturbations of the solitary wave initial data induce modulations of the group parameters. This is consistent with the nonlinear stability result of [1].

Acknowledgement: I am very grateful to the organizers, Prof. S. Albeverio, Prof. Ph. Blanchard and Prof. D. Testard, of the fourth German-French meeting on mathematical physics for inviting me to this stimulating conference. I also would like to thank again Prof. Ph. Blanchard for his constant encouragement during the preparation of my work.

References:

[1] M. Grillakis, J. Shatah and W. Strauss, J. Funct. Anal. 74, 160 (1987)

[2] J. Stubbe, Portugaliae Mathematica, to appear

[3] D.W. McLaughlin and A.C. Scott, Phys. Rev. A18, 11652 (1978)

[4] M.I. Weinstein, Siam J. Math. Anal. 16, 472 (1985)

INDEX

Brownian motion 160
calculus within the differential envelope 83
character of the θ-summable Fredholm module 110
classical closed bosonic strings 191
Connes' operator 114
Cuntz envelope 97
cut-locus sous-Riemannien 17
cyclic boundaries 112
cyclic cocycles 112
cyclic cohomology 112
differential of $\Omega(A)$ 79
directional derivation on Poisson space 64
divergence operator 68
dynamical systems describing plasma phenomena 177
empirical measures 206
energetic stability 236
energy forms in terms of white noise 231
entire cyclic cohomology of a $\mathbb{Z}/2$-graded banach algebra 87
epidemic dynamics as discrete time stochastic processes 36
Euclidean quantization of closed bosonic strings 192
generalized white noise functionals 227
generalized white noise functionals 5
graded kms - functionals 149
gradient operator 68, 229
Hamiltonian system 211
Hida derivatives 5
Hochschild boundary operator 111
Hochschild cochains 111
Hochschild cocycles resp. boundaries 111
Hochschild cohomology 111
Hochschild cohomology group 111
homogeneous chaos 2
Itô's lemma 9
Kolmogoroff's grundbegriffe 162
linear stability 236
Liouville measure 198
Liouville measure 200
localized Polyakov measure 196
magneto-hydrodynamic instabilities 181
Malliavin calculus 63
Malliavin calculus 70
mathematical description of a plasma 175
moderate interaction 206
nondegeneracy condition 213
nonlinear (orbital) stability 236
pathspace measure for closed bosonic strings 194
physical description of a plasma 173
propagation of chaos 207
random graph epidemics 34
reaction-diffusion models 205
solitary wave 235
standard epidemiological models 28
stochastic integral in white noise calculus 6
supertrace 141
twist mapping theorem 221
white noise 2
white noise test functionals 5
Wick-ordering 3
Zekri algebra 97